Physical Diagnosis and Rehabilitation Technologies

Physical Diagnosis and Rehabilitation Technologies

Editors

Tao Liu
João Ferreira

MDPI • Basel • Beijing • Wuhan • Barcelona • Belgrade • Manchester • Tokyo • Cluj • Tianjin

Editors
Tao Liu
School of Mechanical Engineering
Zhejiang University
Hangzhou
China

João Ferreira
Electrical Engineering Department
Superior Institute of Engineering of Coimbra
Coimbra
Portugal

Editorial Office
MDPI
St. Alban-Anlage 66
4052 Basel, Switzerland

This is a reprint of articles from the Special Issue published online in the open access journal *Electronics* (ISSN 2079-9292) (available at: www.mdpi.com/journal/electronics/special_issues/rehabilitation_technologies).

For citation purposes, cite each article independently as indicated on the article page online and as indicated below:

LastName, A.A.; LastName, B.B.; LastName, C.C. Article Title. *Journal Name* **Year**, *Volume Number*, Page Range.

ISBN 978-3-0365-5672-7 (Hbk)
ISBN 978-3-0365-5671-0 (PDF)

Contents

About the Editors

Tao Liu

Tao Liu achieved an M. Eng. degree in Mechanical Engineering at the Harbin Institute of Technology, Harbin, China, in 2003, and a Doctorate degree in Engineering at Kochi University of Technology, Kochi, Japan, in 2006. He was an Assistant Professor in the Department of Intelligent Mechanical Systems Engineering, Kochi University of Technology, Japan, from 2009 to 2013. He is currently a Professor of the State Key Laboratory of Fluid Power and Mechatronic Systems, School of Mechanical Engineering, Zhejiang University, China.

Dr Liu is also the inventor of one Japan patent concerning wearable sensors for gait analysis, which have been commercialized. He was a recipient of the Japan Society of Mechanical Engineers Encouragement Prize (2010). In 2013, he was awarded the Chinese Recruitment Program of Global Youth Experts. His current research interests include wearable sensor systems, rehabilitation robots, biomechanics, and human motion analysis.

João Ferreira

João P. Ferreira achieved his BSc degree in Electrical Engineering in 1999, Master in Industrial Automation in 2002, and PhD in Instrumentation and Control in 2010 at the University of Coimbra. Currently, he is a Coordinator Professor of the Electrical Engineering Department at the Superior Institute of Engineering of Coimbra, Coordinator of the specialization course of Industrial Automation, Robotics and Maintenance, and Researcher at the Institute of Systems and Robotics of the University of Coimbra. He has coordinated and participated in several funding projects in the area of humanoid and medical rehabilitation robotics, with more than 100 scientific publications in international journals/conferences, over 300 participations as a reviewer of scientific manuscripts, and he also holds a national patent (n°. 108143). His research interests include robotics, humanoid robots, human gait, rehabilitation robotics, artificial intelligence, and their applications.

electronics

Editorial

Editorial for the Special Issue on Physical Diagnosis and Rehabilitation Technologies

Tao Liu [1,*] and João Paulo Morais Ferreira [2,*]

1 State Key Laboratory of Fluid Power and Mechatronic Systems, School of Mechanical Engineering, Zhejiang University, Hangzhou 310027, China
2 Electrical Engineering Department, Superior Institute of Engineering of Coimbra, 3030-199 Coimbra, Portugal
* Correspondence: liutao@zju.edu.cn (T.L.); ferreira@isec.pt (J.P.M.F.)

Citation: Liu, T.; Ferreira, J.P.M. Editorial for the Special Issue on Physical Diagnosis and Rehabilitation Technologies. *Electronics* **2022**, *11*, 2247. https://doi.org/10.3390/electronics11142247

Received: 13 July 2022
Accepted: 15 July 2022
Published: 18 July 2022

Publisher's Note: MDPI stays neutral with regard to jurisdictional claims in published maps and institutional affiliations.

Recently, physical diagnosis and human motion analysis have become active research topics in bioelectronics, and they have a broad range of applications, such as pathology detection, rehabilitation, prosthesis design, biometric identification, and humanoid robotic locomotion. Clinical human motion analysis methods aim to provide an objective means of quantifying the severity of pathology. A set of pathology-related human motion disorders have been identified and can be used to support diagnosis and the development of new assistive and rehabilitation technologies. This Special Issue in *Electronics*, titled "Physical Diagnosis and Rehabilitation Technologies", compiles some of the recent research accomplishments in the field of robotics and sensors for human assistance. It consists of 10 papers, which cover rehabilitation robots, human–computer interaction, and sensor and data augmentation, including two review papers. These papers can be categorized into four groups as follows:

(1) Rehabilitation robot: Li et al. [1] proposed an upper-limb rehabilitation training system based on an fNIRS-BCI system. The paper mainly focuses on the analysis and research of the cerebral blood oxygen signal in the system, and gradually extends the analysis and recognition method of the movement intention by using the cerebral blood oxygen signal to implement the actual brain–computer interface system. Some crucial technologies and typical prototypes of active intelligent rehabilitation and assistance systems for gait training are introduced in [2]. The limitations, challenges, and future directions, in terms of gait measurement and intention recognition, gait rehabilitation evaluation, and gait training control strategies, are also discussed. Han et al. [3] reviewed rehabilitation exoskeletons in terms of the overall design, driving unit, intention perception, compliant control, and efficiency validation. They also discussed the complexity and coupling of the human–machine integration systems, and wanted to guide the design of lower-limb rehabilitation exoskeleton systems for elderly and disabled patients. Shi et al. [4] developed a control strategy based on torque estimation and made it responsible for the intention understanding and motion servo of this customized system. Gao et al. [5] provided a dual-armed robotic puncture scheme to assist surgeons. The system was divided into an ultrasound scanning arm and a puncture arm. The robotic arms were designed with a compliant positioning function and master–slave control function.

(2) Human–computer interaction: Based on the results of the users' spatial controllability, Wu et al. [6] proposed two interaction techniques (non-visual selection and a spatial gesture recognition technique for surgery) and four spatial partitioning strategies for human–computer interaction designers, which can improve the users' spatial controllability. To facilitate further developments in flexible display interactive technology, Yin et al. [7] introduced a FlexSheet that can simulate the deformation environment.

(3) Sensor: Han et al. [8] presented a wearable PFC oxygen saturation measurement system using dual-wavelength, functional, near-infrared spectroscopy. The system

was designed for user-friendly wearing, with the advantages of comfort, convenience, portability, and affordability. A novel solid–liquid mixture pressure-sensing module is proposed in [9]. A flexible film with unique liquid-filled structures greatly reduces the pulse measurement error caused by sensor misalignment. The device is expected to provide a new solution for continuous wearable BP monitoring.

(4) Data augmentation: An integrated modeling approach incorporating a war trauma severity scoring algorithm (WTSS) and deep neural networks (DNN) is proposed in [10]. The experimental results verified that the proposed approach surpassed the traditional manual generation methods, achieved a prediction accuracy of 84.43%, and realized large-scale and credible war-trauma data augmentation.

Funding: This research was funded by Open Fund of the State Key Laboratory of Fluid Power and Mechatronic Systems: GZKF-202101.

Acknowledgments: We would like to thank all the authors for the papers they submitted to this Special Issue. We would also like to acknowledge all the reviewers for their careful and timely reviews to help improve the quality of this Special Issue.

Conflicts of Interest: The authors declare no conflict of interest.

References

1. Li, C.; Xu, Y.; He, L.; Zhu, Y.; Kuang, S.; Sun, L. Research on fNIRS Recognition Method of Upper Limb Movement Intention. *Electronics* **2021**, *10*, 1239. [CrossRef]
2. Han, Y.; Liu, C.; Zhang, B.; Zhang, N.; Wang, S.; Han, M.; Ferreira, J.P.; Liu, T.; Zhang, X. Measurement, Evaluation, and Control of Active Intelligent Gait Training Systems—Analysis of the Current State of the Art. *Electronics* **2022**, *11*, 1633. [CrossRef]
3. Wang, T.; Zhang, B.; Liu, C.; Liu, T.; Han, Y.; Wang, S.; Ferreira, J.P.; Dong, W.; Zhang, X. A Review on the Rehabilitation Exoskeletons for the Lower Limbs of the Elderly and the Disabled. *Electronics* **2022**, *11*, 388. [CrossRef]
4. Shi, Y.; Dong, W.; Lin, W.; He, L.; Wang, X.; Li, P.; Gao, Y. Human Joint Torque Estimation Based on Mechanomyography for Upper Extremity Exosuit. *Electronics* **2022**, *11*, 1335. [CrossRef]
5. Gao, Y.; Liu, X.; Zhang, X.; Zhou, Z.; Jiang, W.; Chen, L.; Liu, Z.; Wu, D.; Dong, W. A Dual-Armed Robotic Puncture System: Design, Implementation and Preliminary Tests. *Electronics* **2022**, *11*, 740. [CrossRef]
6. Wu, H.; Han, Y.; Zhou, Y.; Zhang, X.; Yin, J.; Wang, S. Investigation of Input Modalities Based on a Spatial Region Array for Hand-Gesture Interfaces. *Electronics* **2021**, *10*, 3078. [CrossRef]
7. Yin, J.; Bai, S.; Han, Y.; Zhang, X.; Deng, S.; Wang, S. The Study of Bending and Twisting Input Modalities in Deformable Interfaces. *Electronics* **2021**, *10*, 2991. [CrossRef]
8. Han, Y.; Zhai, Q.; Yu, Y.; Wang, S.; Liu, T. A Wearable Prefrontal Cortex Oxygen Saturation Measurement System Based on Near Infrared Spectroscopy. *Electronics* **2022**, *11*, 1971. [CrossRef]
9. Zhan, B.; Yang, C.; Xie, F.; Hu, L.; Liu, W.; Fu, X. An Alignment-Free Sensing Module for Noninvasive Radial Artery Blood Pressure Measurement. *Electronics* **2021**, *10*, 2896. [CrossRef]
10. Yin, J.; Zhao, P.; Zhang, Y.; Han, Y.; Wang, S. A Data Augmentation Method for War Trauma Using the War Trauma Severity Score and Deep Neural Networks. *Electronics* **2021**, *10*, 2657. [CrossRef]

electronics

Article

A Wearable Prefrontal Cortex Oxygen Saturation Measurement System Based on Near Infrared Spectroscopy

Yi Han [1,2,†], Qian Zhai [1,†], Yinkai Yu [1], Shuoyu Wang [2] and Tao Liu [1,*]

[1] State Key Laboratory of Fluid Power and Mechatronic Systems, School of Mechanical Engineering, Zhejiang University, Hangzhou 310027, China; 258012g@gs.kochi-tech.ac.jp (Y.H.); 11825031@zju.edu.cn (Q.Z.); 21960582@zju.edu.cn (Y.Y.)
[2] Department of Intelligent Mechanical Systems Engineering, Kochi University of Technology, Kochi 782-8502, Japan; wang.shuoyu@kochi-tech.ac.jp
* Correspondence: liutao@zju.edu.cn; Tel.: +86-0571-87951314 (ext. 6221)
† These authors contributed equally to this work.

Abstract: The measurement of blood oxygen saturation in the prefrontal cortex (PFC), especially during sleep, is of great significance for clinical research. Herein, we present a wearable PFC oxygen saturation measurement system using dual-wavelength functional near-infrared spectroscopy. The system is well designed for user-friendly donning and has the advantages of comfort, convenience, portability, and affordability. The performance of the proposed system is investigated by the calibration and experimental results. The wearable system has demonstrated great potential to conduct the physiological monitoring of PFC, and it can be widely deployed in daily life.

Keywords: near infrared spectroscopy; wearable; portable; prefrontal cortex oxygen saturation

Citation: Han, Y.; Zhai, Q.; Yu, Y.; Wang, S.; Liu, T. A Wearable Prefrontal Cortex Oxygen Saturation Measurement System Based on Near Infrared Spectroscopy. *Electronics* **2022**, *11*, 1971. https://doi.org/10.3390/electronics11131971

Academic Editor: J.-C. Chiao

Received: 14 May 2022
Accepted: 15 June 2022
Published: 24 June 2022

1. Introduction

The prefrontal cortex (PFC), responsible for complex thinking and decision-making, as well as emotional regulation, is fundamental to the physiological activities of humans [1]. Nightly sleep restores the brain and is correlated with physical repairs and memory consolidation. A sufficient and stable oxygen supply to cerebral cortex tissue is essential for the relaxation of the brain [2]. Several diseases cause various local microvascular or respiratory modifications, resulting in changes of tissue oxygen saturation [3,4]. Subtle alterations in sleep quality result in consequential next-day changes in emotion or intentness [5]. Therefore, the accurate measurement of blood oxygen saturation in the PFC tissue is of great significance for clinical research. In this way, physiological information can be acquired, and then corresponding effective treatments can be taken in time [6].

In the 1870s, Jobsis [7] first obtained blood oxygen changes and deep physiological information in the heads of animals through near infrared light. The absorption of light by the main absorbers (i.e., oxyhemoglobin (HbO_2) and deoxyhemoglobin (Hb)) in the blood varies with the oxygen saturation. Hence, the absorbed components of the light reflected from the human tissue can be analyzed to obtain the blood oxygen saturation [8]. With the advantages of portability, affordability, low susceptibility to noise, and moderate temporal resolution compared with the medical polysomnography (PSG) procedure, functional near-infrared spectroscopy (fNIRS) based technologies have attracted increasing attention [9]. Oniz et al. investigated the refreshing property of sleep in terms of sleep stages using fNIRS to measure PFC hemodynamics [10]. Nguyen et al. investigated the brain's functional connectivity in rest and sleep states using fNIRS [11]. Ahn et al. combined fNIRS with multimodal EEG/ECG/EOG data to quantify mental fatigue during performed simulated driving under two different conditions (well-rested and sleep-deprived) [12]. The previous research works have demonstrated that some of the activities and states during sleep are relevant to the blood oxygen changes of PFC. The fNIRS-based technologies enable the non-invasive sensing of these physiological phenomena. Combined with other modalities, fNIRS can aid in characterizing and understanding the complex phenomena and be applied in sleep monitoring. Nevertheless, sleep disorders, such as obstructive sleep apnea, significantly deteriorate the cerebral oxygen supply to the brain. Little research has illuminated the effects of respiratory modifications during sleep using fNIRS. There are also some well-known commercial products, such as Somanetics INVOS [13], CASMED FORESIGHT [14], Hamamatsu NIRO [15], and Shimadzu OM-100, which have been applied in clinical monitoring. These devices are complex, expensive, and restricted to hospitals. The past decade has witnessed an increasing expectation for in-home health management [16]. However, the fNIRS-based research on PFC physiology, especially during sleep, is primary, and the current measurement systems are not suitable for daily monitoring. To monitor the physiological activities of PFC continuously and conveniently, the aim of this research is to design a novel fNIRS-based wearable system, validate the performance of monitoring PFC oxygen saturation, and investigate its potential for sensing-related activities, especially during sleep. The system is miniaturized, user-friendly, affordable, and convenient for applications in non-hospital facilities.

The rest of the paper is organized as follows. In Section 2, the analytical model and the measurement system are presented in detail. Section 3 describes the experimental setup and protocols. In Section 4, the results are illustrated and discussed. Finally, a brief conclusion is drawn.

2. Measurement System Description

2.1. Analytical Model

Assuming that the absorption rate and scattering coefficient of biological tissue are constant, the blood oxygen saturation is highly relevant to the decay of light intensity due to the absorption of Hb and HbO_2 [17]. The modified Lambert–Beer law describes the quantitative relationship between the light absorption and the characteristics of the tissue:

$$B^{\lambda} = \log(\frac{I_0}{I}) = (\alpha^{\lambda}_{HbO_2} C_{HbO_2} + \alpha^{\lambda}_{Hb} C_{Hb}) \cdot DPF \cdot r + G, \tag{1}$$

where B is the attenuation of light intensity, I_0 and I are the incident light intensity and emitted light intensity, respectively, α is the absorption coefficient of the chromophores at a specific wavelength, λ is the wavelength of the light, C_{HbO_2} and C_{Hb} are the concentrations of HbO_2 and Hb, DPF denotes the differential path factor (e.g., 6.53 in our application [18]), r is the distance between light source–detector pairs, and G is a geometry-dependent factor representing the intensity loss caused by scattering. When the concentrations of HbO_2 and Hb change, the attenuation of light intensity changes homogeneously:

$$\Delta B^{\lambda} = \log(\frac{I_0}{I}) = (\alpha^{\lambda}_{HbO_2} \Delta C_{HbO_2} + \alpha^{\lambda}_{Hb} \Delta C_{Hb}) \cdot DPF \cdot r, \tag{2}$$

If the absorption changes are measured at two wavelengths, λ_1 and λ_2, the concentration changes of HbO_2 and Hb can be computed as

$$\begin{aligned} \Delta C_{HbO_2} &= \frac{\Delta B^{\lambda_1} \alpha^{\lambda_2}_{Hb} - \Delta B^{\lambda_2} \alpha^{\lambda_1}_{Hb}}{DPF \cdot r \cdot (\alpha^{\lambda_2}_{Hb} \alpha^{\lambda_1}_{HbO_2} - \alpha^{\lambda_1}_{Hb} \alpha^{\lambda_2}_{HbO_2})}, \\ \Delta C_{Hb} &= \frac{\Delta B^{\lambda_1} \alpha^{\lambda_2}_{HbO_2} - \Delta B^{\lambda_2} \alpha^{\lambda_1}_{HbO_2}}{DPF \cdot r \cdot (\alpha^{\lambda_1}_{Hb} \alpha^{\lambda_2}_{HbO_2} - \alpha^{\lambda_2}_{Hb} \alpha^{\lambda_1}_{HbO_2})}, \end{aligned} \tag{3}$$

The blood oxygen saturation rSO_2 can be calculated as

$$rSO_2 = \frac{C_{HbO_2}}{C_{HbO_2} + C_{Hb}} = \frac{\alpha^{\lambda_1}_{Hb} - \alpha^{\lambda_2}_{Hb} \frac{B^{\lambda_1}}{B^{\lambda_2}}}{\frac{B^{\lambda_1}}{B^{\lambda_2}} (\alpha^{\lambda_2}_{HbO_2} - \alpha^{\lambda_2}_{Hb}) - (\alpha^{\lambda_1}_{HbO_2} - \alpha^{\lambda_1}_{Hb})}. \tag{4}$$

Since light with a wavelength longer than 900 nm is easily absorbed by water and light with a wavelength shorter than 700 nm is excessively absorbed by hemoglobin, the dual wavelengths of 760 nm and 850 nm are selected for measurement.

2.2. *System Structure*

The structure of the fNIRS-based measurement system is shown in Figure 1. The system consists of the micro-processor unit (MPU), infrared emission module, sampling module, Bluetooth transmission module, and graphical user interface (GUI). The infrared emission module is fixed to the forehead of subjects and controls the emission of near-infrared light. Then, the sampling module detects the intensity of the received light. Finally, the blood oxygen saturations at different positions in the PFC are calculated by the MPU. The results will be transmitted to the GUI via Bluetooth.

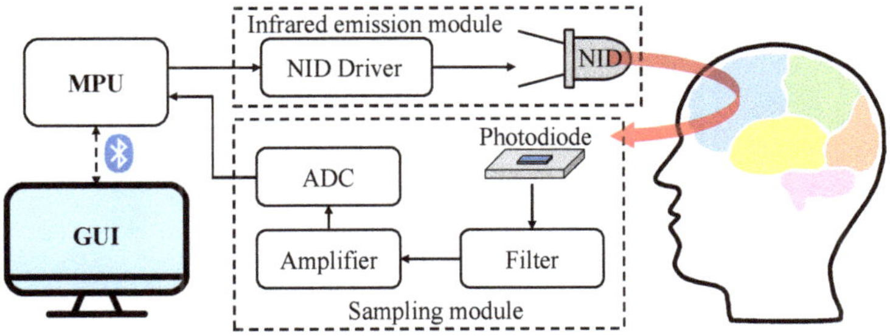

Figure 1. The schematic diagram of the designed system structure.

The infrared emission module consists of near-infrared diodes (NID) and an NID driver. In consideration of light intensity and penetration depth, the module utilizes the diodes (e.g., W760/850-O4L) that integrate sources of dual wavelengths (i.e., 760 nm and 850 nm). It has the advantages of a small size, low cost, and embedability. The MPU and the constant current driver (e.g., MBI5168) with multiple outputs control the emission timing of multiple NIDs. The sampling module consists of photodiodes, a filter, an amplifier, and an analog-to-digital-converter (ADC). The photodiodes (e.g., OPT101) receive the emitted near-infrared light and convert light signals into electrical signals. They have high sensitivity within the wavelength from 700 nm to 900 nm. Since the change of the PFC oxygen saturation is relatively low compared with external noises, low-pass filters are employed to eliminate high-frequency interference. Then, the filtered signals are magnified by the amplifier (e.g., AD623) and are converted into digital signals by the ADC channels. Moreover, we select STM32 for timing control and oxygen saturation calculation, HC-05 for Bluetooth transmission, and SPX3819 for power supply.

2.3. *Wearable Design*

Figure 2 illustrates the distribution of the NID and photodiodes in one probe module. One NID is placed in the center of four photodiodes, and the distance between each NID–photodiode pair is approximately 28 mm. Hence, the near-infrared light can transmit through the deeper tissue layers.

As shown in Figure 3, two probe modules are employed in the system, and the distance between the NIDs is approximately 60 mm. To minimize the interference resulting from the environment or movement, it is essential to block out the natural light from the detection area and ensure a tight connection between the sensor and the forehead. In each probe module, a baffle is fixed in the back as a shelter. The connection of probe modules adopts the hinge structure, and silicone is used for the contact between modules and forehead to better fit the various subjects and improve wearing comfort.

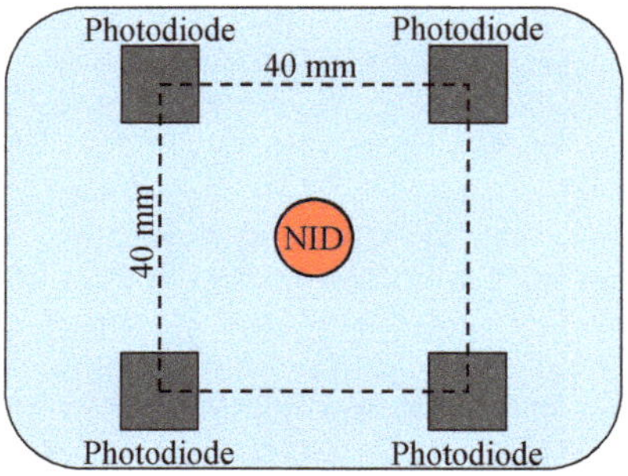

Figure 2. The distribution diagram of the probes.

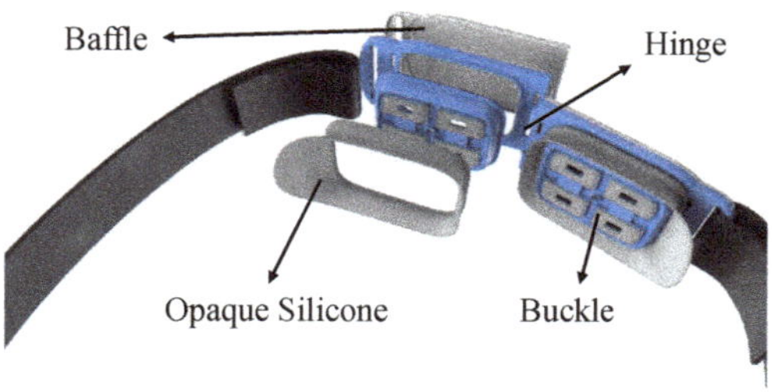

Figure 3. The wearable design of the measurement system.

3. Experiments

Firstly, the optical integrating sphere model as illustrated in Figure 4 is exploited to simulate human tissue [19]. The emitted near infrared light that transmits through the sphere model has the same optical power as the incident light. By adjusting the light filter, the model can imitate the attenuation effect of human tissue with a different oxygen saturation and realize the multi-point calibration of oxygen saturation measurement.

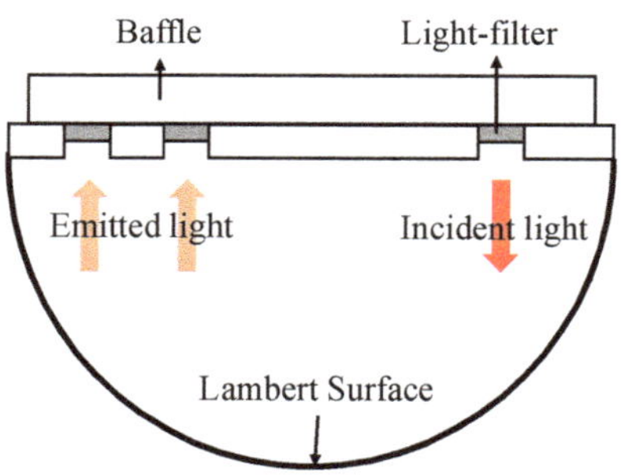

Figure 4. Diagram of the optical integrating sphere model.

In order to verify the performance and reliability of the system, a forearm arterial occlusion experiment, motor imagery experiment, and sleep experiment were performed. Participants aged from 20 to 25 years old were recruited for the experiments. The participants had no major diseases, and the blood pressure of each participant was normal and stable.

As shown in Figure 5a, each participant wore an inflatable airbag in the forearm arterial occlusion experiment. When the inflated airbag pressed the arm, the concentrations of HbO_2 and Hb changed correspondingly. The steps of the forearm arterial occlusion experiment were implemented as follows:

1. Let the participant sit upright in the chair and place his right arm parallel on the table;

2. Fix the measurement system to the front side of the arm and fix the airbag to the upper arm of the participant;
3. Turn on the measurement system when the participant relaxes;
4. Fifteen seconds later, inflate the airbag to press the upper arm;
5. Fifteen seconds later, release the airbag;
6. Turn off the measurement system.

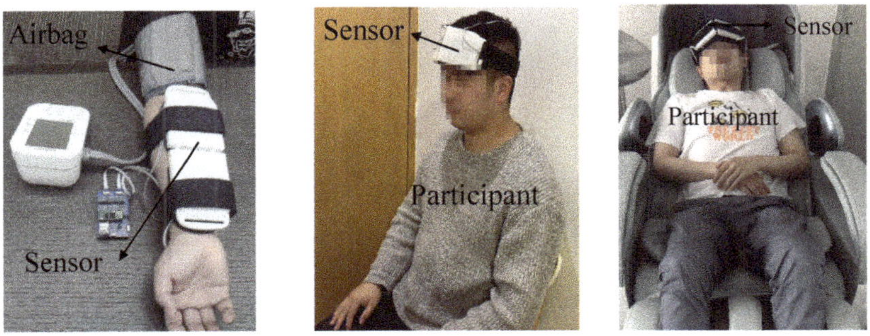

(**a**) Forearm Block Experiment (**b**) Motor Imagery Experiment (**c**) Sleep Experiment

Figure 5. (**a**) Each participant wore the measurement system and an inflatable airbag on their arm in the forearm arterial occlusion experiment. (**b**) Each participant wore the measurement system on their forehead in the motor imagery experiment. (**c**) Each participant lay in a pod and wore the measurement system on their forehead in the sleep experiment.

The PFC is responsible for complex thinking and decision-making. The blood oxygen saturation of this brain region will change with physiological activities. To detect the movement intention, each participant wore the measurement system on their forehead, as shown in Figure 5b in the motor imagery experiment. The steps were implemented as follows:

1. Let the participant sit in the chair and fix the measurement system onto their forehead;
2. Turn on the system and let the participant close their eyes throughout the experiment;
3. Ten seconds later, ask the participant to imagine walking upstairs;
4. Forty seconds later, let the subject stop imagining and relax;
5. Turn off the measurement system.

Different biological processes through the sleep cycle probably affect breathing or the muscles. Deep breathing or sleep apnea will result in a change of the blood oxygen saturation in the PFC. As shown in Figure 5c, the participants lay in a nap pod and were asked to breathe normally, breathe deeply, and hold their breath. Clinical medical staffs were consulted to help subjects to conduct the experiments and record the events. The steps were implemented as follows:

1. Let the participant lie in the nap pod and fix the measurement system onto their forehead;
2. Turn on the system and let the participant close their eyes throughout the experiment;
3. Fifteen seconds later, ask the participant to breathe deeply or hold their breath;
4. Thirteen seconds later, let the subject breathe normally;
5. Turn off the measurement system.

4. Results

Figure 6 shows the calibration results where the triangles represent the measured values, the red line represents the ideal line, and the black line represents the fitted line. The measurement range of the oxygen saturation is between 0.5–0.75, which is the same as human tissue. The correlation coefficient between the fitted line and ideal line is about 0.98, and the result demonstrates the accuracy of the system.

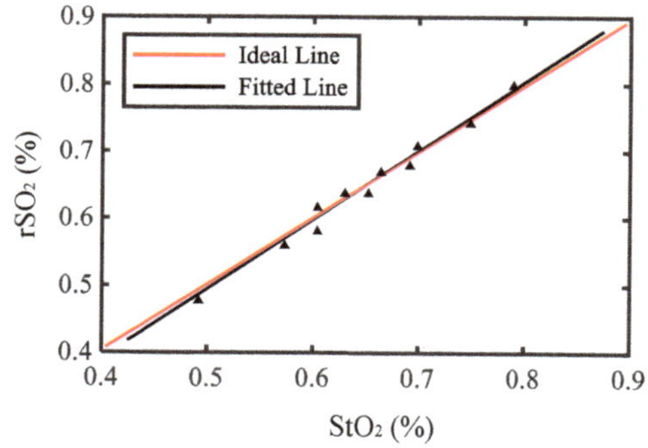

Figure 6. The calibration result of the system.

Figure 7 shows one example of the measurement result in the forearm arterial occlusion experiment. At the beginning, the concentrations of HbO_2 and Hb remained stable when the participant stayed relaxed. When the airbag was inflated and the forearm arterial was occluded, blood flow to the forearm was restricted as well as the oxygen. Hence, HbO_2 in the forearm was partially converted into Hb when the oxygen was consumed. Consistently, the measured concentration of HbO_2 increased and the concentration of Hb decreased in Figure 7. After the airbag was released, the two concentrations returned to the normal state.

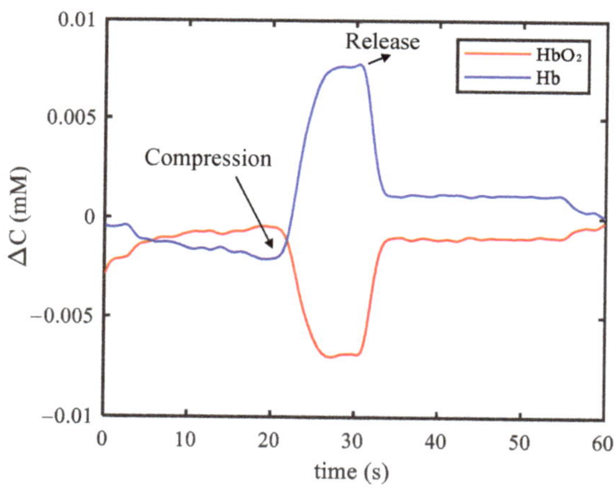

Figure 7. Example of results in the forearm arterial occlusion experiment.

Figure 8 shows one example of the measurement result in the motor imagery experiment. The concentrations of HbO_2 and Hb were unchanged in the relaxing condition. When the participant started to imagine walking upstairs, the blood flow in PFC and the transmission of oxygen increased. It is also illustrated as in Figure 8 that the concentration of HbO_2 increased sharply and the concentration of Hb also had a slight increase.

Figure 9a shows one example of the measurement result in the deep breath experiment. The deep breath improved the blood supply to the brain. When the participant started to breathe deeply, the concentration of HbO_2 increased rapidly and the concentration of Hb decreased. Finally, the concentrations returned to the normal level. Figure 9b shows one example of the measurement result in the apnea experiment. Holding breath increased the intrathoracic pressure, resulting in restricted blood flow. When the participant started to hold their breath, the concentration of HbO_2 decreased, and the concentration of Hb had a slight increase as shown in Figure 9c. When the participant breathed normally again, the concentrations slowly returned to the normal level.

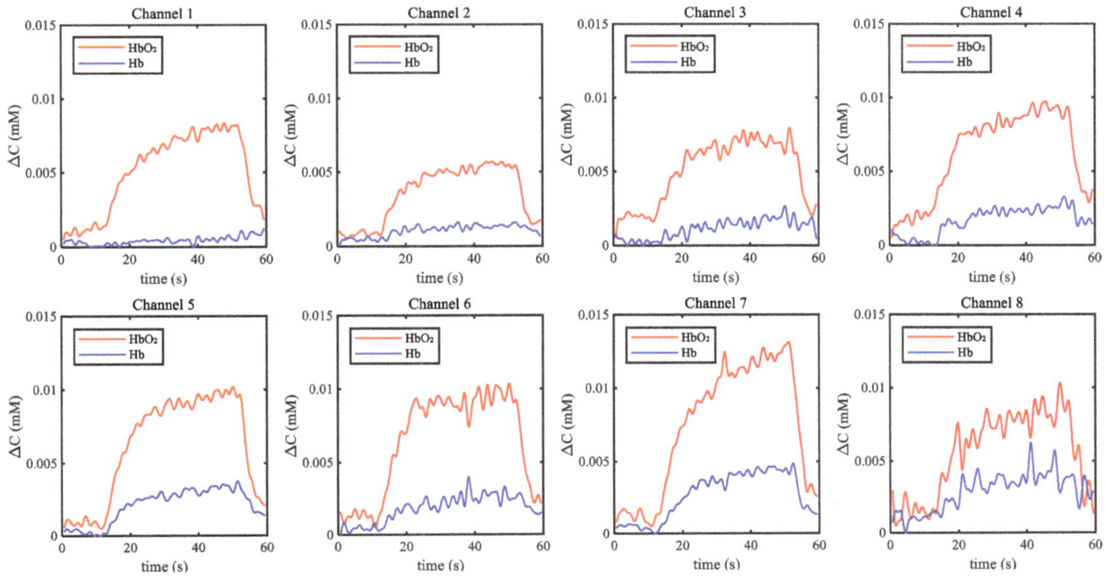

Figure 8. Example of results in the motor imagery experiment.

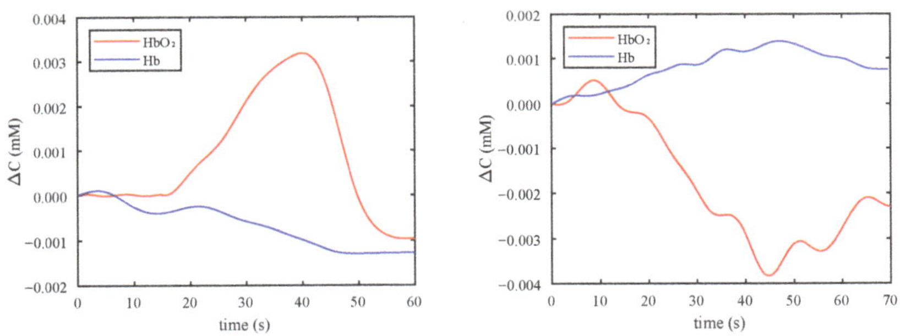

(**a**) Deep Breath Experiment (**b**) Apnea Experiment

Figure 9. (**a**) Example of results in the deep breathing experiment. (**b**) Example of results in the apnea experiment.

5. Discussion

The calibration and experimental results demonstrate that the changes of the concentrations of HbO_2 and Hb in PFC can be measured accurately by the system. The correlation coefficient of 0.98 in the calibration is comparable with a coefficient larger than 0.95 in [8]. The physiological phenomenon in the forearm arterial occlusion experiment and motor imagery experiment are consistent with results in [8,20], respectively. The measurement system has dimensions of $17.5 \times 7 \times 3.5$ cm and weighs around 600 g, which is smaller and lighter compared with the system that has dimensions of $18.5 \times 7.5 \times 7$ cm and a weight of around 1200 g in [21]. Moreover, conventional near-infrared blood oxygen measurement devices have some disadvantages, such as a complex structure, large size, high cost, and wearing discomfort. The proposed sensor unit employs a hinge in its structure to enable adaptive deformation and soft silicone as a material for flexible contact with the forehead. With the miniaturization and lightweight design, the wearable system is comfortable, convenient, and not expensive. It has great potential to get rid of site restriction and be promoted to non-hospital facilities.

Currently, the proposed system also has some limitations. First, the system is verified based on healthy subjects only, and we do not focus on a specific population, such as patients with sleep apnea. Second, the sensor units may generate pressure on the forehead to ensure the stable contact. Third, the human-computer interface and measurement region are preliminarily designed for non-clinical experimental tests. For future work, the software interface will be further optimized for clinical application. Then the system can be employed to implement experiments of patients and record over-night sleep events. We will explore methods to detect the physiological activities including sleep apnea.

6. Conclusions

A wearable PFC oxygen saturation measurement system based on fNIRS is proposed in this research. The system is comfortable, easy-to-use, portable, and low-cost. The performance of the wearable blood oxygen saturation measurement system is investigated by the calibration accuracy and experimental results. The changes of the hemoglobin concentrations in PFC can be measured accurately by the system. It is demonstrated that the proposed system has great potential to conduct the physiological monitoring of the PFC, and it can be widely deployed in non-hospital facilities and in daily life.

Author Contributions: Conceptualization, T.L. and S.W.; methodology, Y.H. and Y.Y.; software, Y.H. and Y.Y.; validation, Y.Y.; formal analysis, Q.Z.; investigation, Y.H. and Q.Z.; resources, T.L. and S.W.; data curation, Q.Z.; writing—original draft preparation, Q.Z.; writing—review and editing, Q.Z.; visualization, Y.Y.; supervision, T.L.; project administration, Y.H.; funding acquisition, T.L. All authors have read and agreed to the published version of the manuscript.

Funding: This research was funded in part by the NSFC Grant No. 52175033 and No. U21A20120, in part by Zhejiang Provincial Natural Science Foundation of China under Grant LZ20E050002, in part by Zhejiang Provincial Key Research and Development Program under Grant 2021C03051, 2021C03G2013248, and 2019C03080, and in part by Hangzhou Leading Innovation and Entrepreneurship Team under Grant 2018TD06.

Institutional Review Board Statement: The study was approved by the Medical Ethics Committee of Zhejiang Hospital (Approval Letter NO: AF/SC-06/04.2).

Conflicts of Interest: The authors declare no conflict of interest.

References

1. Leopold, D.; Phillips, T. Chapter 7—The importance of a good night's sleep. In *A Prescription for Healthy Living*; Academic Press: Cambridge, MA, USA, 2021; pp. 93–102.
2. Chisholm, K.; Ida, K.; Davies, A.; Papkovsky, D.; Singer, M.; Dyson, A.; Tachtsidis, I.; Duchen, M.; Smith, K. In Vivo Imaging of Flavoprotein Fluorescence During Hypoxia Reveals the Importance of Direct Arterial Oxygen Supply to Cerebral Cortex Tissue. *Adv. Exp. Med. Biol.* **2016**, *876*, 233. [PubMed]
3. Sircan-Kucuksayan, A.; Uyuklu, M.; Canpolat, M. Measuring tissue oxygen saturation using NIR spectroscopy. In *Proceedings of the Biophotonics: Photonic Solutions for Better Health Care IV. International Society for Optics and Photonics*; SPIE: Bellingham, WA, USA, 2014; Volume 9129, pp. 369–376.
4. Benjafield, A.; Ayas, N.; Eastwood, P.; Heinzer, R.; Ip, M.; Morrell, M.; Nunez, C.; Patel, S.; Penzel, T.; Pépin, J.L.; et al. Estimation of the global prevalence and burden of obstructive sleep apnoea: A literature-based analysis. *Lancet Respir. Med.* **2019**, *7*, 687–698. [CrossRef]
5. Ben Simon, E.; Rossi, A.; Harvey, A.; Walker, M. Overanxious and underslept. *Nat. Hum. Behav.* **2020**, *4*, 100–110. [CrossRef] [PubMed]
6. Kurth, C.; Levy, W.; Mccann, J. Near-Infrared Spectroscopy Cerebral Oxygen Saturation Thresholds for Hypoxia Ischemia in Piglets. *J. Cereb. Blood Flow Metab.* **2002**, *22*, 335–341. [CrossRef] [PubMed]
7. Jöbsis, F. Noninvasive, infrared monitoring of cerebral and myocardial oxygen sufficiency and circulatory parameters. *Science* **1977**, *198*, 1264–1267. [CrossRef] [PubMed]
8. Zhu, J.; Mou, J.; Liu, J. Absolute measurement of tissue oxygen saturation with NIRS. In Proceedings of the BIBE 2018 International Conference on Biological Information and Biomedical Engineering, Shanghai, China, 6–8 July 2018; pp. 1–5.
9. Sorensen, L.C.; Leung, T.S.; Greisen, G. Comparison of cerebral oxygen saturation in premature infants by near-infrared spatially resolved spectroscopy: Observations on probe-dependent bias. *J. Biomed. Opt.* **2008**, *13*, 064013. [CrossRef] [PubMed]
10. Oniz, A.; Inanc, G.; Taşlıca, S.; Guducu, C.; Ozgoren, M. Sleep Is a Refreshing Process: An fNIRS Study. *Front. Hum. Neurosci.* **2019**, *13*, 160. [CrossRef] [PubMed]
11. Nguyen, T.; Babawale, O.; Kim, T.; Jo, H.; Liu, H.; Kim, J. Exploring brain functional connectivity in rest and sleep states: A fNIRS study. *Sci. Rep.* **2018**, *8*, 16144. [CrossRef] [PubMed]

12. Ahn, S.; Nguyen, T.; Jang, H.; Kim, J.; Jun, S. Exploring Neuro-Physiological Correlates of Drivers' Mental Fatigue Caused by Sleep Deprivation Using Simultaneous EEG, ECG, and fNIRS Data. *Front. Hum. Neurosci.* **2016**, *10*, 219. [CrossRef]
13. Marik; Paul, E. Regional carbon dioxide monitoring to assess the adequacy of tissue perfusion. *Curr. Opin. Crit. Care* **2005**, *11*, 245. [CrossRef]
14. Nagdyman, N.; Ewert, P.; Peters, B.; Miera, O.; Fleck, T.; Schmitt, B. Comparison of different near-infrared spectroscopic cerebral oxygenation indices with central venous and jugular venous oxygenation saturation in children. *Paediatr. Anaesth.* **2008**, *18*, 160–166. [CrossRef] [PubMed]
15. Gagnon, R.; Macnab, A.; Gagnon, F.; Blackstock, D.; LeBlanc, J. Comparison of Two Spatially Resolved NIRS Oxygenation Indices. *J. Clin. Monit. Comput.* **2002**, *17*, 385–391. [CrossRef] [PubMed]
16. Maswadi, K.; Ghani, N.B.A.; Hamid, S.B. Systematic Literature Review of Smart Home Monitoring Technologies Based on IoT for the Elderly. *IEEE Access* **2020**, *8*, 92244–92261. [CrossRef]
17. Chance, B.; Zhuang, Z.; UnAh, C.; Alter, C.; Lipton, L. Cognition-activated low-frequency modulation of light absorption in human brain. *Proc. Natl. Acad. Sci. USA* **1993**, *90*, 3770–3774. [CrossRef] [PubMed]
18. Duncan, A.; Meek, J.H.; Clemence, M.; Elwell, C.E.; Tyszczuk, L.; Cope, M.; Delpy, D. Optical pathlength measurements on adult head, calf and forearm and the head of the newborn infant using phase resolved optical spectroscopy. *Phys. Med. Biol.* **1995**, *40*, 295–304. [CrossRef] [PubMed]
19. Bashkatov, A.; Genina, E.; Tuchin, V. Optical properties of skin, subcutaneous, and muscle tissues: A review. *J. Innov. Opt. Health Sci.* **2011**, *04*, 9–38. [CrossRef]
20. Sitaram, R.; Zhang, H.; Guan, C.; Thulasidas, M.; Hoshi, Y.; Ishikawa, A.; Shimizu, K.; Birbaumer, N. Temporal classification of multichannel near-infrared spectroscopy signals of motor imagery for developing a brain–computer interface. *NeuroImage* **2007**, *34*, 1416–1427. [CrossRef] [PubMed]
21. Everdell, N.; Airantzis, D.; Kolvya, C.; Suzuki, T.; Elwell, C. A portable wireless near-infrared spatially resolved spectroscopy system for use on brain and muscle. *Med. Eng. Phys.* **2013**, *35*, 1692–1697. [CrossRef] [PubMed]

electronics

Review

Measurement, Evaluation, and Control of Active Intelligent Gait Training Systems—Analysis of the Current State of the Art

Yi Han [1,2,†], Chenhao Liu [1,†], Bin Zhang [1], Ning Zhang [3], Shuoyu Wang [2], Meimei Han [4], João P. Ferreira [5], Tao Liu [1] and Xiufeng Zhang [3,*]

1 State Key Laboratory of Fluid Power and Mechatronic Systems, School of Mechanical Engineering, Zhejiang University, Hangzhou 310027, China; 258012g@gs.kochi-tech.ac.jp (Y.H.); 12125056@zju.edu.cn (C.L.); zjuzhangbin@zju.edu.cn (B.Z.); liutao@zju.edu.cn (T.L.)
2 Department of Intelligent Mechanical Systems Engineering, Kochi University of Technology 185 Miyanokuchi, Tosayamada-cho, Kami-city 782-8502, Japan; wang.shuoyu@kochi-tech.ac.jp
3 Key Laboratory of Rehabilitation Technical Aids Technology and System of the Ministry of Civil Affairs, National Research Center for Rehabilitation Technical Aids, Beijing 100176, China; zhangning@nrcrta.cn
4 Zhejiang Fuzhi Science and Technology Innovation Co., Ltd., Hangzhou 310027, China; mmhan@zju.edu.cn
5 Institute of Superior of Engineering of Coimbra, Quinta da Nora, 3030-199 Coimbra, Portugal; ferreira@mail.isec.pt
* Correspondence: zhangxiufeng@nrcrta.cn
† These authors contributed equally to this work.

Abstract: Gait recognition and rehabilitation has been a research hotspot in recent years due to its importance to medical care and elderly care. Active intelligent rehabilitation and assistance systems for lower limbs integrates mechanical design, sensing technology, intelligent control, and robotics technology, and is one of the effective ways to resolve the above problems. In this review, crucial technologies and typical prototypes of active intelligent rehabilitation and assistance systems for gait training are introduced. The limitations, challenges, and future directions in terms of gait measurement and intention recognition, gait rehabilitation evaluation, and gait training control strategies are discussed. To address the core problems of the sensing, evaluation and control technology of the active intelligent gait training systems, the possible future research directions are proposed. Firstly, different sensing methods need to be proposed for the decoding of human movement intention. Secondly, the human walking ability evaluation models will be developed by integrating the clinical knowledge and lower limb movement data. Lastly, the personalized gait training strategy for collaborative control of human–machine systems needs to be implemented in the clinical applications.

Keywords: rehabilitation and assistance system; lower limbs; intention recognition; gait training; gait evaluation; human–machine interaction control strategy

Citation: Han, Y.; Liu, C.; Zhang, B.; Zhang, N.; Wang, S.; Han, M.; Ferreira, J.P.; Liu, T.; Zhang, X. Measurement, Evaluation, and Control of Active Intelligent Gait Training Systems—Analysis of the Current State of the Art. *Electronics* **2022**, *11*, 1633. https://doi.org/10.3390/electronics11101633

Academic Editor: Dong-Joo Kim

Received: 12 April 2022
Accepted: 19 May 2022
Published: 20 May 2022

1. Introduction

Walking is one of the most common behaviors in human daily life, and the ability to walk is an important factor for human beings to live independently. However, neurological diseases such as stroke sequelae and Parkinson's disease can lead to impairment of human motor function and decline in walking ability [1], which can seriously affect the quality of life and health of patients. The World Health Organization survey shows that the incidence of stroke in China ranks first in the world, and stroke is characterized by high incidence rate, high disability rate, high mortality rate, and high recurrence rate [2]. According to the report of the National Bureau of Statistics of China, the elderly population in China will reach 267 million, accounting for 18.9% of the national population in 2022. The accelerated process of aging has increased the number of people suffering from neurological diseases, and the conflict with the lack of medical resources has become a serious problem in the health care system [3]. At present, human beings cannot break the laws of nature to prevent the decline in their own motor functions, and many injuries to the body's motor function are irreversible. It has become one of the urgent problems in society to help the elderly or patients overcome movement disorders, restore their walking function, and improve their daily living ability.

Active intelligent gait training systems are robotic devices that actively interact with human lower limbs to provide support and assistance for the body's motor function. State-of-the-art rehabilitation training walkers or robotic systems mainly have problems such as poor gait adaptability, inability to quantify and feedback rehabilitation effect, single training strategy, and limited sensor usage environment [4–12]. Facing major national needs and the main economic battlefield, it is of great significance to develop intelligent rehabilitation equipment to realize gait perception, evaluation, and feedback in the home environment, and to help rehabilitation physiotherapists to assist patients in restoring motor function. The gait training system is a large area of research which integrates mechanical design, sensing technology, intelligent control, and robotics technology. At the forefront of the research of intelligent gait training systems and evaluation methods, there are three important basic scientific problems to be solved, which include: (1) the measurement of lower limb movement and the prediction of movement intention, (2) the construction of a walking ability evaluation model based on clinical knowledge and lower limb movement data, and (3) the formulation of personalized gait training strategy of collaborative control of human–machine system. Therefore, the key words "gait measurement and intention recognition", "gait evaluation", and "gait training control strategy" were used in the literature review. Additionally, this review not only selected publications that directly describe or introduce any gait training system, but also retains those publications that focus on any of the three basic scientific problems mentioned above.

In this review, the current active intelligent gait training systems are investigated and discussed from three perspectives, in accordance with three critical scientific problems put forward above, which are measurement and prediction of lower limb movement, evaluation of the effect of gait rehabilitation, and the control strategy of gait training. The main limitations and challenges are then discussed, and potential future directions of intelligent gait training systems are put forward.

2. Human Gait Measurement and Intention Recognition

2.1. Gait Movement Measurement

The active intelligent gait training systems have the ability to monitor patient's movement in real time [13]. At present, human's body movements are mainly measured through the fixed force platform and optical motion capture system [14–16] in the gait laboratory, or multiple movement and force sensors worn on the limb [17–19]. The former is highly accurate but limited by the measurement environment, and the latter may interfere with the normal human movement.

The current main human movement measurement methods used in gait training systems are shown in Figure 1. Gait motion measurement techniques used in each of the included studies [14–36] and their characteristics are shown in Table 1. Vision-based methods are one of the important methods for monitoring the posture and movement of the human body and have a wide range of applications [20–23]. Based on the image global joint summation problem or the hierarchical detection fusion problem, deep learning methods have been widely studied for the estimation of human pose [24,25]. However, visual methods have problems such as clothing occlusion, dark environment, high system complexity, difficult installation, and privacy issues, and there are limitations in actual human–machine coordinated movement. The wearable sensing system of human body dynamics analysis consists of multiple sensors, including gyroscopes, pressure sensors, angle sensors, inertial sensors, etc., but it has difficulties in obtaining displacement and relative poses from human to machine. The radio frequency (RF) signal-based method can use the data characteristics of the human body and its motion in the radar image to measure the three-dimensional relative pose and radial velocity [35]. The latest research [36] shows that it has obvious advantages in solving problems such as occlusion and three-dimensional reconstruction, but at present it still needs in-depth research on issues such as decoupling RF signals of human and machine movement, fusion understanding based on physical models and data, and generalized measurement of abnormal gait. Therefore, it is necessary to study a new type of non-contact sensing technology solution, combining the kinematics information of the lower limbs and plantar pressure detection to form a composite information perception system to accurately predict the movement trend of the patient's lower limbs and use it to evaluate the patient's health and athletic ability.

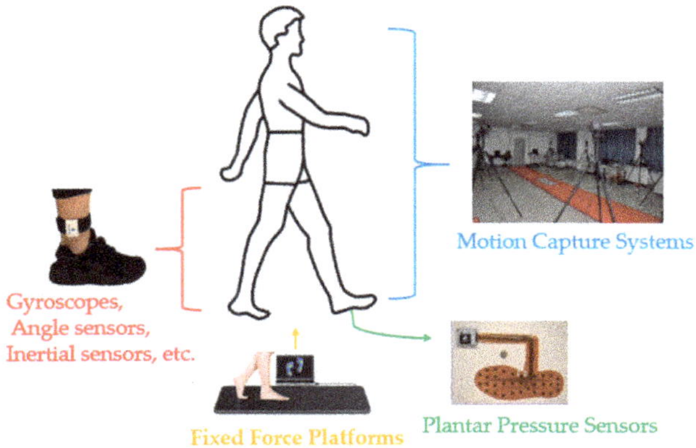

Figure 1. State-of-the-art motion measurement techniques used in gait training systems.

Table 1. Gait motion measurement techniques used in each of the included studies and their characteristics.

Study	Sensing Techniques	Advantages	Challenges
Jensen et al. [14] Xie et al. [20] Bao et al. [21] Steinert et al. [22] Tran et al. [23] Toshev et al. [24]	One camera Three noncontact cameras Pupil Labs eye tracking system 2D smartphone camera Seven Kinect sensors(cameras) Images taken by a camera	Motion capture systems represented by Vicon are currently regarded as the "gold standard" for motion capture by their high accuracy	Greatly affected by the environment, high system complexity, difficult to install, and privacy issues
Yang et al. [35] Zhao et al. [36]	Radio-frequency identification (RFID) tags Radio frequency (RF) signals from RF-Avatar	Measure in the presence of occlusions, baggy clothes, and bad lighting conditions	Decoupling of RF signals of human–machine coordinated movement
Veilleux et al. [15]	Six large force platforms	No image information will be left, and user privacy will not be violated	Only in the laboratory
Zeng et al. [16] Mazhar et al. [17] Trkov et al. [18] Li et al. [19] Schicketmueller et al. [26,29]	Smart sensor shoes A flex sensor on a leather shoe inertial sensors on lower limbs Designed strain gauge on leg Inertial measurement units	Unrestricted use environment, simple to use, user privacy will not be violated	May intervene with the normal motion, difficulties in obtaining displacement and relative pose of human–machine
Martini et al. [27] Wang et al. [30] Bae et al. [31] Livolsi et al. [32] Bae et al. [33] Chen et al. [34]	Embedded joint angle sensors Foot pressure sensor and IMU Force sensors in the foot plates Hip encoders, pressure-insoles Inertial measurement units A single IMU	Unrestricted use environment, simple to use, user privacy will not be violated	May intervene with the normal motion, difficulties in obtaining displacement and relative pose of human–machine

2.2. Movement Intention Recognition

For patients with impaired motor function but can still move and do some daily activity, the willingness to actively participate in gait training is especially important in the rehabilitation process [37]. Clinical studies have shown that active involvement of patients in the rehabilitation training is more effective in the neurological reconstruction and motor function recovery. Therefore, as an important input information of active intelligent rehabilitation and assistant robotics device, human lower limb movement intention needs to be captured in real time.

The current typical intention recognition methods used in gait training systems and the statistics of recent research studies using each method are shown in Table 2. The neuro-rehabilitation training robotics devices should show "transparency" in the patient's walking assistance process, that is, reduce the intervention of the patient's active gait as much as possible [38], and the key lies in the understanding and prediction of the patient's movement intention. Current intention recognition methods are mainly based on bioelectric signals and motion signals. The electroencephalogram (EEG) signals are the overall reflection of the movement intentions in brain [39], and have the shortest latency, but it has a low signal-to-noise ratio, easily interfered by noise [40,41]. The Electromyographic (EMG) signals reflect the state of muscle activation and the feedback control based on EMG signal can effectively improve the human–machine coordination in rehabilitation training [42–45], but it has strong ambiguity and is affected by factors such as surface contact status, muscle displacement, and muscle fatigue [46]. The intention recognition method based on motion signal mainly uses kinematic signals such as position, angle, and speed, and kinetic signals such as interaction force/torque, which has high reliability, robustness, and accuracy [47–50]. Xu et al. [51] proposed a compliance control algorithm for walking-aid robots based on multi-sensor fusion, which allows the robot to obey human movement by recognizing user intentions. Esteban et al. [52] also carried out related research, using EMG signals and Artificial Neural Network (ANN) algorithms to recognize human walking intention and proposed a robotic knee exoskeleton for assistance and rehabilitation. Wu et al. [53] put forward a coordinated control strategy based on human–machine interaction and the principle of minimum interference. However, the information of human motion and force is the result of the movement, with a certain time lag between the motion intent. Therefore, in response to the active adjustment needs of human-in-the-loop control, it is necessary to study new motion perception systems and intention prediction models with self-learning capabilities, and to improve the stability, synergy and adaptability of human–machine collaboration based on active intention feedback.

Table 2. Intent recognition method used in each of the included studies and their characteristics.

Intent Recognition Methods	Study	Characteristic
Electroencephalogram (EEG) signal method	Liu et al. [39]	High accuracy: 80.16 ± 5.44%
	Engemann et al. [40]	The best model depends on noise
	Bi et al. [41]	To recognize intention under the attended and distracted states
Electromyographic (EMG) signal method	Zhuang et al. [42]	Proved to be better than interaction-torque based method
	Zhang et al. [43]	Back Propagation (BP) neural network was used
	Xie et al. [44]	General regression neural network optimized by golden section algorithm was used
	Rabe et al. [45]	Anterior sonomyography sensor fusion with surface EMG
	Fougner et al. [46]	3.8~18% average classification error due to muscle fatigue
	Mora-Tola et al. [52]	Artificial Neural Network (ANN) algorithms were used

Table 2. *Cont.*

Intent Recognition Methods	Study	Characteristic
Kinetic signals method	Guo et al. [47]	A robot dynamics model including the active force of human was established, and contact force was used to analyze intention
	Pinheiro et al. [50]	The interaction torque's direction and magnitude were both used
	Xu et al. [51]	A compliance control algorithm based on intent was proposed
	Wu et al. [53]	A minimal-intervention-based admittance control strategy was developed
Kinematic signals method	Gong et al. [48]	Two IMUs and an imbedded BPNN-based algorithm were used
	Zhu et al. [49]	Recognition accuracy rate can reach 97.64%

3. Evaluation of Gait Rehabilitation

Clinical gait analysis and evaluation is of great significance in active intelligent gait training systems. Quantitative analysis methods based on sensor data are important methods for gait rehabilitation evaluation. An increasing number of researchers in physical therapy, bioengineering, neurology and rehabilitation have been participating in this field of study. In the early research studies, gait analysis and evaluation usually took the form of scales, such as the Fugl-Meyer exercise scale [54]. According to the scale, medical staff perform the diagnosis and evaluation of motor function, the monitoring of disease progression, and the evaluation of curative effect. The result of evaluation is often affected by a large number of subjective and inaccurately measurable parameters in the clinical scale [55].

Table 3 shows number of gait evaluation studies, which sensors and features were used in each research [56–77], and the real-time of gait evaluation. Gait parameters are usually used to assist medical staff in diagnosis, rating and scoring of motor function, monitoring the progress of the patient's condition, and evaluating curative effect. Gait measurement equipment such as motion capture systems and wearable inertial sensors have been widely used in clinical practice. Some researchers used the gait parameters measured by these large systems to predict Parkinson's diagnosis and Hoehn-Yahr (H-Y) classification [67,68]. There are also researchers who used the changes in gait parameters before and after the patient receives treatment and training to evaluate the treatment effect [69]. Caramia et al. [70] used eight inertial measurement units placed on the lower extremities and trunk to estimate several gait parameters such as step length, stride speed, etc. and extract features from them to distinguish between healthy people and patients with H-Y grades 1 to 3 in order to achieve diagnosis and grade prediction. However, there are problems such as inconvenient use of sensing equipment, lack of clinical significance of data features, difficulty in matching the scale, and an incomplete assessing system. Wang et al. [71] carried out preliminary research based on clinical needs, using as few human sensor measurement data as possible, and using nonlinear data classification methods to achieve quantitative evaluation of dyskinesias in patients with abnormal gait. Skvortsov et al. [72] also investigated the feasibility of gait analysis and walking function evaluation based on the stance phase of stroke patients using biofeedback technology.

Muscle synergy theory describes a potential neuromuscular control mechanism of vertebrate limb movement [73]. According to the muscle synergy theory, nerves do not control a certain muscle alone, but recruit muscles on the spinal cord to form muscle groups, that is, muscle synergy. The muscles in the same muscle synergy are activated at the same time. Compared with controlling each muscle individually, using one control signal to activate multiple muscles theoretically provides a simplified system. Numerous experimental research results support this theory [74,75]. Studies have shown that muscle activation during motor tasks can be described in terms of low-dimensional control that reflects muscle synergy. The downward commands of the nervous system to the musculoskeletal system are manifested in muscle synergy, which is reflected in muscle activation through spinal cord circuits or reflexes, thereby forming a force in the musculoskeletal system, driving the musculoskeletal system to move and producing specific actions.

Table 3. Sensors and features used in gait evaluation method in each of the included studies.

Study	Sensors Used	Features Used in Gait Evaluation	Real-Time
Anaya-Reyes et al. [56]	Vicon MX T20	Step phase durations and cadence	–
Alberto et al. [57]	3-D motion capture system	Stride width and gait velocity	–
Ma et al. [58]	Three Force Sense Resistors	Knee and hip joints and FSRs data	–
Chomiak et al. [59]	Ambulosono system	Step length, distance traveled, velocity, and cadence	+
Tran et al. [60]	A motion capture system and four force sensors	Center of mass, the center of pressure, and step parameters	–
Park et al. [61]	Force sensors	Angles, active force, and resistive force	–
Sconza et al. [62]	Dynamometer	Knee extensor strength, double-time support, and step length ratio	–
Wang et al. [65]	3-D motion capture system	Cadence and single stance time	–
Tamburella et al. [66]	Angle sensors	Gait speed	–
Wahid et al. [67]	8-camera video motion analysis system	Stride length, step length, and double support time	–
Rehman et al. [68]	GAITRite instrument	Step velocity and step length	–
Carlotta et al. [70]	Inertial measurement units (IMU)	Step length, step time, and stride speed	–
Wang et al. [71]	Inertial measurement units (IMU)	Right spatial-temporal and kinematic gait parameters	+
Skvortsov et al. [72]	Neurosens inertial sensors	Knee and hip joint range of motion	–
Cheung et al. [74]	EMG sensors	EMG signals (muscle activity)	–
Safavynia et al. [75]			–
Longatelli et al. [77]			–
Rinaldi et al. [76]	EMG sensors, Vicon, and force platform	Both gait parameters and muscle activity	–
Seo et al. [78]	EMG sensors and IMU		–

Abbreviations: + real-time evaluation; – off-line evaluation.

Based on the muscle synergy theory, many research studies have been carried out to diagnose gait disorders and neurological diseases by measuring the activation state of lower limb muscles during walking [76–80]. However, the existing methods for measuring muscles exercise have drawbacks. On the one hand, Surface Electromyography (sEMG) signal measurement has limitations which include the lack of ability to test the deep muscles, the easily interfered EMG sensors, and the difficulty for the extraction process of the EMG signal envelope to accurately demodulate the neural excitation when the motor neuron action potential is generated. On the other hand, Indwelling Electromyography (iEMG) causes a certain degree of damage to human muscles, which is not suitable for long-term exercise detection with multiple measurements. At the same time, the existing simulation software is generally based on a variety of rule constraints such as muscle force-length relationship constraints, muscle force and joint motion coupling constraints, etc., and optimization theories such as minimizing physiological consumption. However, according to the results of human motion modeling and analysis by related researchers [79,80], in patients with gait disorders, it is often difficult to meet the above constraints due to nerve-muscle-skeletal damage, and the dynamic representations such as joint torque are affected by motion compensation under the condition of external load changes.

4. Control Strategy of Gait Training Systems

Traditional walking devices, such as crutches, walkers, wheelchairs, etc., are mostly passive devices, which cannot solve the problem of coordinated dynamic training of body and lower limb muscle when the elderly and patients walk [81]. On the contrary, active intelligent mobility assistance devices interact physically with the human body, as well as coordinated movement, to provide support and assistance for the body's motor function and help the body maintain and restore its motor function to the greatest extent [82,83]. The typical control diagram of active intelligent gait training systems is shown in Figure 2.

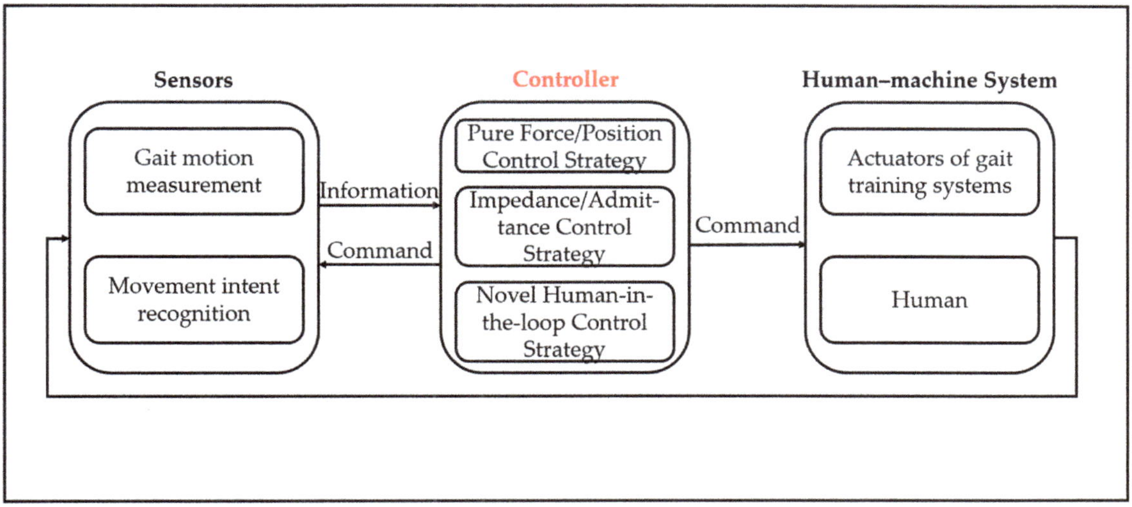

Figure 2. Typical control diagram of gait training systems.

A walker is a walking rehabilitation assistive device used to assist users in standing and walking activities which can effectively help users improve their walking ability and is of great significance to a large number of disabled or elderly people. Based on this, intelligent walking rehabilitation assistive robotics devices effectively use the current rapid development of technology to help users break through the original limitations of walking ability to a certain extent and improve their mobility to meet their daily needs. These technologies include mechanical design technology, embedded system technology, sensing and detection technology, automatic control technology, motor control technology, microelectronics technology, interface technology, and software programming. Table 4 shows several gait training systems and their control strategies from recent studies [84–102]. Pure force/position control means the gait training systems make corresponding operations by detecting human gait events, but do not care about the information of human–machine interaction, while systems with impedance/admittance control strategy use the human–machine interaction information such as interaction force/torque and relative position. Novel human-in-the-loop control represents a large number of control strategies that both recognize the motion intention of humans and detect the human–machine interaction information, described as 'human-in-the-loop' because the information of the human body takes part in both the input and feedback of the closed-loop control. Yu et al. [84] developed an intelligent three-wheeled mobility aid, which is equipped with infrared sensors and laser rangefinders to ensure human–machine–environmental intelligent interaction in motion. Tao et al. [85] studied the intelligent mobility assistance rehabilitation training device for the needs of standing and gait rehabilitation. A standing support and gait training system that maximizes the patient's own rehabilitation exercise ability was developed by using the pressure sensor on the sole of the foot to detect the user's balance or falling state and feeding back the human lower limb joint and muscle force to a load-reducing suspension system. Zhao et al. [88] developed a gait rehabilitation robot to improve the safety and availability of rehabilitation training for patients. A built-in-robot camera was used to obtain leg movement data, and the knee angle was estimated by a New-type ESMF algorithm to deal with the problem of the brief disappearance of the marker point in the field of view.

Table 4. Gait training systems and control strategies of each included.

Device Name	Control Strategy	References
Gait rehabilitation device	Open-loop position control based on GRFs	Tao et al. [85]
Novel Robotic Walker	Position control based on gait event	Ye et al. [86]
Gait Training Robot	Force control method based on gait event	Miyake et al. [87]
Walking Assist Robot	Position control based on fall detection	Zhao et al. [88]
Hybrid Rehabilitation Robot	Manually changed training modes and speed control	Kim et al. [91]
C-ALEX	Open-loop force control base on gait event	Hidayah et al. [96]
Gait Assist Robot	Training mode switch based on task and gait feature	Scheidig et al. [99]
Lower-Limb Exoskeleton	Speed control based on gait parameters	Ma et al. [100]
Intelligent Mobility Aid	Admittance-based mobility controller	Yu et al. [84]
Rotational Orthosis	Admittance control of the ankle mechanism	Mu et al. [89]
Gait rehabilitation robot	Adaptive admittance control based on interactive force	Guo et al. [90]
Improved rotational orthosis	Admittance control based on ankle force	Mu et al. [92]
Clinic gait training robot	Admittance control based on relative position	Shunki et al. [93]
2-DOF Exoskeleton	Admittance control based on interactive force	Chen et al. [101]
Robot Assisted Gait Training	Assist-as-needed Control based on prediction	Zhang et al. [94]
GAREX	Logic compliance adaptation and assist-as-needed	Zhong et al. [95]
Ankle Robotic Orthosis	Assist-as-needed Impedance Control Strategy	Lopes et al. [97]
Biofeedback Exoskeleton	Speed control based on predicted user response	Zhang et al. [98]
AGoRA	Closed-loop control based on intention and gait feature	Mayag et al. [102]

Functional Electrical Stimulation (FES) is a method of applying low-frequency pulsed current or amplifying it through signal-current conversion and then sending it into the human body to produce immediate effects, artificially causing movement in humans who are paralyzed by damage to the central nervous system. Recently, a large number of research studies proposed robotic systems for gait rehabilitation based on FES method [103–105]. Studies have proven that, combined with FES, the assistive torque required of the gait training systems can be reduced and the muscle strength and joint range of motion of the human body can be improved. However, due to the use of electrode pads, this rehabilitation strategy still has problems such as the inability to stimulate deeper muscles or the trauma of electrode implantation in sEMG and iEMG in Section 3.

Locomat is a robotic gait training system. It is used for gait training for patients with abnormal gait caused by brain injury, spinal injury, neurological injury, muscle injury, and orthopedic diseases, and to improve the motor ability of patients with neurological diseases. In the first few generations of prototypes, Locomat also used the common impedance control based on torque feedback [106], but in the latest generations of Locomat Pro, novel control strategy such as automatic gait-pattern adaptation and path control strategy are applied. Locomat Pro can also perform diagnostic evaluation of patients' gait and there are many cases of clinical application [107–109]. However, it is difficult for such a bulky and expensive product to enter millions of households, and the compliance of the control can still be improved. For patients who have lost their mobility due to nerve damage, how to fully mobilize the patient's own movement intention instead of "passive walking" so as to achieve the treatment of nerve injury diseases is a difficult point in the study of the intelligent gait rehabilitation training systems.

5. Limitations and Challenges

Rehabilitation and training of gait is a current research hot spot. From the systematic analysis of the current research status of the active intelligent gait training systems, it is not difficult to see that there are still key issues in terms of sensing, evaluation, and control. Key technologies such as the decoupling of radio frequency signals of human–machine coordinated movement, the understanding based on fusing physical models and gait data, and the generalized measurement of abnormal gait are in urgent need of breakthroughs. To be specific, when capturing patients' motion using RF signals, both the wearable gait training device and the human body reflect RF signals. That makes the decoupling of the return signals from the two very important, and it is also the limitation of current research studies. The methods of lower limb movement analysis and movement intention prediction based on radio frequency principles need to be further studied. As the current intention recognition based on EEG signals is easily interfered by noise, the EMG signals-based method has strong ambiguity. Moreover, the intention recognition based on kinematic or kinetic signals has a long latency.

As a mobility aid for gait rehabilitation and training, if the evaluation criteria for the rehabilitation effect are difficult to define and the efficacy cannot be guaranteed, it will be difficult to meet the diverse and personalized needs of users. The current clinical scale for gait analysis and evaluation relies on the subjective assessment of the doctor and the self-perception of the patient. In addition, the existing sensing data features lack clinical significance and are difficult to correspond to the scale, and the evaluation system is inadequate. How to quantitatively evaluate the effect of gait training with multi-dimensional information still needs in-depth research.

A crucial problem of control of the active intelligent rehabilitation assistance devices for lower limbs is that it needs to allow the users to spontaneously participate in motion, which is of great importance for patients with nerve injury. However, the current gait rehabilitation training systems have difficulty accurately recognizing the user's movement intentions to make corresponding assistance strategies. As users' requirements for comfort and safety continue to increase, the human-in-the-loop control, with information of human body taking part in both the input and feedback of the controller, is receiving increasing attention. However, due to the difficulty of quickly and accurately identifying the user's intent, the research studies on human–machine cooperative intelligent control for personalized gait rehabilitation training is still too preliminary.

6. Future Directions

To address the core problems in the sensing, evaluation, and control technology of the active intelligent gait training systems, the following possible future research directions are proposed. We believe that the key is to focus on scientific issues such as the decoding of lower limb movement intention based on the principle of radio frequency, the construction of a walking ability evaluation model combined with clinical knowledge base and lower limb movement data. and the personalized gait training strategy for collaborative control of human–machine systems.

Among the many methods of detecting and sensing human lower limb movement, the method based on the radio frequency signal is relatively preliminary, but it has obvious advantages and broad prospects. A new type of non-contact motion sensing method based on the principle of millimeter wave echo reflection needs to be studied. For instance, a non-contact small radio frequency sensor such as a millimeter wave radar first needs to be developed. Using the signal features generated by human motion on the radar image and Doppler signal spectrogram as target features, similar to Daniel et al. [110], and using the space occupancy status and motion frequency shift information contained in the frequency characteristic data of the range view as input, the features in the input data are encoded by the convolutional neural network (CNN) method, and an estimator is generated to output the joint position and motion information of the object [111]. By combining the real-time data with the models of the kinematics and dynamics of human lower limbs, the human lower limb movement may be predicted. In conclusion, it is of important scientific significance to study a new non-contact sensing principle and the method of model-driven and data-driven fusion, to integrate the characteristics of different information dimensions, to build a more concise, fast, and accurate online decoding model of composite information for patient's gait training, and to predict patient's movement intentions.

Based on the knowledge of rehabilitation medicine, combined with the results of motion recognition and prediction, the evaluation model of gait rehabilitation training effects needs to be established, and the method of generating personalized rehabilitation training prescriptions needs to be studied. Based on the extracted non-steady-state motion signals of the lower limbs, the time–frequency characteristics of the vital signs signals such as EMG signals can be analyzed. In order to evaluate the movement synergy of the healthy and abnormal limbs of the human body, the mechanism of human muscle synergy needs to be further studied. Combining the lower extremity musculoskeletal model with the static optimization algorithm to calculate the muscle activation degree during human walking, the evaluation method of the muscle coordination degree on the lower extremity muscle movement coordination ability of the patient's exercise training can be studied. Finally, the evaluation of gait training effects for different ages and different training stages will be realized.

The workflow of an ideal gait training system should be as follows: based on the evaluation of walking ability and the needs of gait rehabilitation training, the movement mode of gait training can be determined. The corresponding human motion intention and motion reference trajectory are obtained through the non-contact motion sensing system. After this, the desired motion trajectory of the gait training system is generated. Combined with the motion intention of the lower limbs of the human body and some simple control methods, the gait training system will flexibly assist the patient to complete the desired action. All in all, the key to the formulation of control strategies is gait evaluation and intention recognition, while obscure and sophisticated control theory is secondary. By studying the collaborative control method of the gait rehabilitation training system and the patient, based on principal component analysis, multiple regression, and neural network, the association model between gait data and clinical evaluation can be constructed, and a personalized gait training strategy with multi-layer, and cooperative closed-loop control of "human in the loop" can be designed. Based on this, carrying out research on the collaborative control of human–machine systems based on personalized rehabilitation strategies, evaluating the perception and control performance of the gait training system, and generating clinical evaluation reports on the effects of rehabilitation training have important academic significance and extensive clinical application value.

Funding: This work was supported in part by the National Natural Science Foundation of China under Grant 52175033, U21A20120 and U1913601; the Zhejiang Provincial Natural Science Foundation of China under Grant No. LZ20E050002; Open Fund of the State Key Laboratory of Fluid Power and Mechatronic Systems: GZKF-202101; DongGuan Innovative Research Team Program (2020607202006).

Institutional Review Board Statement: Not applicable.

Informed Consent Statement: Not applicable.

Data Availability Statement: Not applicable.

Conflicts of Interest: The authors declare no conflict of interest.

References

1. Li, N.; Yang, J.; Feng, X.; Zhang, J.; Yang, X.; Zhang, Z. A summary of 30 years' research on risk factors of stroke mortality in China. *Chin. J. Behav. Med. Brain Sci.* **2017**, *26*, 765–768.
2. Wang, L.; Liu, J.; Yang, Y.; Peng, B.; Wang, Y. China Stroke Prevention still faces Great Challenges: China Stroke Prevention Report 2018 Summary. *China Circ. J.* **2019**, *34*, 2.
3. Balaban, B.; Tok, F. Gait Disturbances in Patients With Stroke. *Pmr* **2014**, *6*, 635–642. [CrossRef] [PubMed]
4. Ye, J.; Chen, G.; Liu, Q.; Duan, L.; Wang, C. Gait Phase Estimation for FES Based on Pelvic Movement of a Novel Gait Rehabilitation Robot. In Proceedings of the 2018 IEEE International Conference on Real-time Computing and Robotics (RCAR), Kudahuvadhoo, Maldives, 1–5 August 2018; IEEE: New York, NY, USA, 2018; pp. 532–537.
5. Seo, W.; Kwon, J.H.; Ko, Y.; Hong, D. An Assistive Control Strategy Using Arm Swing Information for 1DoF Hip Exoskeleton for Hemiplegic Gait Rehabilitation. *J. Korean Soc. Precis. Eng.* **2019**, *36*, 419–424. [CrossRef]
6. Qin, T.; Meng, X.; Qiu, J.; Zhu, D.; Zhang, J. Dynamics Analysis of the Human-Machine System of the Assistive Gait Training Robot. In *Intelligent Robotics and Applications, ICIRA*; Springer: Berlin/Heidelberg, Germany, 2019; Volume 11745.
7. Swank, C.; Trammell, M.; Bennett, M. The utilization of an overground robotic exoskeleton for gait training during inpatient rehabilitation-single-center retrospective findings. *Int. J. Rehabil. Res.* **2020**, *43*, 206–213. [CrossRef]
8. Afzal, T.; Tseng, S.; Lincoln, J.; Kern, M. Exoskeleton-assisted Gait Training in Persons With Multiple Sclerosis: A Single-Group Pilot Study. *Arch. Phys. Med. Rehabil.* **2020**, *101*, 599–606. [CrossRef]

9. Hoyer, E.; Opheim, A. Implementing the exoskeleton Ekso GT^{TM} for gait rehabilitation in a stroke unit–feasibility, functional benefits and patient experiences. *Disabil. Rehabil. Assist. Technol.* **2020**. [CrossRef]
10. Wang, F.C.; Li, Y.C.; Wu, K.L.; Chen, P.Y.; Fu, L.C. Online gait detection with an automatic mobile trainer inspired by neuro-developmental treatment. *Sensors* **2020**, *20*, 3389. [CrossRef]
11. Ma, W.; Huang, R.; Chen, Q.; Song, G.; Li, C. Dynamic Movement Primitives based Parametric Gait Model for Lower Limb Exoskeleton. In Proceedings of the 2020 39th Chinese Control Conference (CCC), Shenyang, China, 27–30 July 2020; pp. 3857–3862.
12. Ji, J.; Wang, Y.; Zhang, G.; Lin, Y.; Wang, G. Design and Simulation Analysis of a Robot-Assisted Gait Trainer with the PBWS System. *J. Healthc. Eng.* **2021**, *2021*, 2750936. [CrossRef]
13. Liang, P.; Hou, Z.; Wang, C.; Luo, L.; Wang, W. Rehabilitation-assisted robot and their physical human-computer interaction methods. *J. Autom.* **2018**, *44*, 2000–2010.
14. Jensen, R.R.; Paulsen, R.R.; Larsen, R. Analysis of gait using a treadmill and a time-of-flight camera. In Proceedings of the Workshop on Dynamic 3D Vision, Jena, Germany, 9 September 2009.
15. Veilleux, L.N.; Raison, M.; Rauch, F.; Robert, M.; Ballaz, L. Agreement of spatio-temporal gait parameters between a vertical ground reaction force decomposition algorithm and a motion capture system. *Gait Posture* **2016**, *43*, 257–264. [CrossRef]
16. Zeng, D.; Qu, C.; Ma, T. Research on a gait detection system and recognition algorithm for lower limb exoskeleton robot. *J. Braz. Soc. Mech. Sci. Eng.* **2021**, *43*, 298. [CrossRef]
17. Mazhar, O.; Bari, A.Z.; Faudzi, A. Real-time gait phase detection using wearable sensors. In Proceedings of the Control Conference, Chengdu, China, 27–29 July 2016; IEEE: New York, NY, USA, 2016.
18. Trkov, M.; Chen, K.; Yi, J.; Liu, T. Inertial Sensor-Based Slip Detection in Human Walking. *IEEE Trans. Autom. Sci. Eng.* **2019**, *16*, 1399–1411. [CrossRef]
19. Li, G.; Song, J.; Wang, X.; Lan, F.; Zou, F. Research on Lower Limb Exoskeleton Based on Multi-Sensor Information Mature Technology. In Proceedings of the 2019 IEEE 9th Annual International Conference on CYBER Technology in Automation, Control, and Intelligent Systems (CYBER), Suzhou, China, 29 July–2 August 2019; pp. 808–813.
20. Xie, L.; Huang, L. Wirerope-driven exoskeleton to assist lower-limb rehabilitation of hemiplegic patients by using motion capture. *Assem. Autom.* **2020**, *40*, 48–54. [CrossRef]
21. Bao, W.; Villarreal, D.; Chiao, J. Vision-Based Autonomous Walking in a Lower-Limb Powered Exoskeleton. In Proceedings of the 2020 IEEE 20th International Conference on Bioinformatics and Bioengineering (BIBE), Cincinnati, OH, USA, 26–28 October 2020; pp. 830–834.
22. Steinert, A.; Sattler, I.; Otte, K.; Röhling, H.; Mansow-Model, S.; Müller-Werdan, U. Using New Camera-Based Technologies for Gait Analysis in Older Adults in Comparison to the Established GAITRite System. *Sensors* **2020**, *20*, 125. [CrossRef]
23. Tran, T.-H.; Nguyen, D.T.; Phuong Nguyen, T. Human Posture Classification from Multiple Viewpoints and Application for Fall Detection. In Proceedings of the 2020 IEEE Eighth International Conference on Communications and Electronics (ICCE), Phu Quoc Island, Vietnam, 13–15 January 2021; pp. 262–267.
24. Toshev, A.; Szegedy, C. DeepPose: Human Pose Estimation via Deep Neural Networks. In Proceedings of the 2014 IEEE Conference on Computer Vision and Pattern Recognition, Columbus, OH, USA, 23–28 June 2014.
25. Lee, B.; Kim, J.; Jung, S.-U. Light-weighted Network based Human Pose Estimation for Mobile AR Service. In Proceedings of the 2020 International Conference on Information and Communication Technology Convergence (ICTC), Jeju, Korea, 21–23 October 2020; pp. 1609–1612.
26. Schicketmueller, A.; Rose, G.; Hofmann, M. Feasibility of a Sensor-Based Gait Event Detection Algorithm for Triggering Functional Electrical Stimulation during Robot-Assisted Gait Training. *Sensors* **2019**, *19*, 4804. [CrossRef]
27. Martini, E.; Crea, S.; Parri, A.; Bastiani, L.; Faraguna, U.; McKinney, Z.; Molino-Lova, R.; Pratali, L.; Vitiello, N. Gait training using a robotic hip exoskeleton improves metabolic gait efficiency in the elderly. *Sci Rep.* **2019**, *9*, 7157. [CrossRef]
28. Lee, H.; Afzal, M.R.; Pyo, S.; Yoon, J. A Novel Gait Assistance System Based on an Active Knee Orthosis and a Haptic Cane for Overground Walking. In *Converging Clinical and Engineering Research on Neurorehabilitation III*; Springer: Berlin/Heidelberg, Germany, 2019; Volume 21, pp. 439–443.
29. Schicketmueller, A.; Lamprecht, J.; Hofmann, M.; Sailer, M.; Rose, G. Gait Event Detection for Stroke Patients during Robot-Assisted Gait Training. *Sensors* **2020**, *20*, 3399. [CrossRef]
30. Wang, T.; Pei, X. An untethered cable-driven ankle exoskeleton with plantarflexion-dorsiflexion bidirectional movement assistance. *Front. Inform. Technol. Electron. Eng.* **2020**, *21*, 723–739. [CrossRef]
31. Bae, E.; Park, S.; Moon, Y. A Robotic Gait Training System with Stair-climbing Mode Based on a Unique Exoskeleton Structure with Active Foot Plates. *Int. J. Control Autom. Syst.* **2020**, *18*, 196–205. [CrossRef]
32. Livolsi, C.; Conti, R.; Giovacchini, F.; Vitiello, N.; Crea, S. A Novel Wavelet-Based Gait Segmentation Method for a Portable hip Exoskeleton. *IEEE Trans. Robot.* **2021**, 1–15. [CrossRef]
33. Bae, J.; Lee, J. Design Of A Control System For A Lower-Limb Exoskeleton Rehabilitation Robot With Gait Phase Detection Algorithm Using Inertial Sensor. *Assist. Technol.* **2021**, *33*, 166.
34. Chen, S.; Bangaru, S.S.; Yigit, T.; Trkov, M.; Wang, C.; Yi, J. Real-Time Walking Gait Estimation for Construction Workers using a Single Wearable Inertial Measurement Unit (IMU). In Proceedings of the 2021 IEEE/ASME International Conference on Advanced Intelligent Mechatronics (AIM), Delft, The Netherlands, 12–16 July 2021; pp. 753–758.

35. Yang, C.; Wang, X.; Mao, S. RFID-Pose: Vision-Aided Three-Dimensional Human Pose Estimation With Radio-Frequency Identification. *IEEE Trans. Reliab.* **2021**, *70*, 1218–1231. [CrossRef]
36. Zhao, M.; Liu, Y.; Raghu, A.; Zhao, H.; Katabi, D. Through-Wall Human Mesh Recovery Using Radio Signals. In Proceedings of the 2019 IEEE/CVF International Conference on Computer Vision (ICCV), Seoul, Korea, 27 October–2 November 2019.
37. Meng, W.; Liu, Q.; Zhou, Z.; Ai, Q.; Sheng, B.; Xie, S. Recent development of mechanisms and control strategies for robot-assisted lower limb rehabilitation. *Mechatronics* **2015**, *31*, 132–145. [CrossRef]
38. Zhang, T.; Huang, H. Design and Control of a Series Elastic Actuator With Clutch for Hip Exoskeleton for Precise Assistive Magnitude and Timing Control and Improved Mechanical Safety. *IEEE/ASME Trans. Mechatron.* **2019**, *24*, 2215–2226. [CrossRef]
39. Liu, D.; Chen, W.; Pei, Z.; Wang, J. A brain-controlled lower-limb exoskeleton for human gait training. *Rev. Sci. Instrum.* **2017**, *88*, 104302. [CrossRef]
40. Engemann, D.A.; Gramfort, A. Automated model selection in covariance estimation and spatial whitening of MEG and EEG signals. *Neuroimage* **2015**, *108*, 328–342. [CrossRef]
41. Bi, L.; Xia, S.; Fei, W. Hierarchical Decoding Model of Upper Limb Movement Intention From EEG Signals Based on Attention State Estimation. *IEEE Trans. Neural Syst. Rehabil. Eng.* **2021**, *29*, 2008–2016. [CrossRef]
42. Zhuang, Y.; Yao, S.; Ma, C.; Song, R. Admittance Control Based on EMG-Driven Musculoskeletal Model Improves the Human–Robot Synchronization. *IEEE Trans. Ind. Inform.* **2019**, *15*, 1211–1218. [CrossRef]
43. Zhang, L.; Ma, Y.; Wang, C.; Yan, Z.; Wu, X. A Method for Arm Motions Classification and A Lower-limb Exoskeleton Control Based on sEMG signals. In Proceedings of the 2019 IEEE 4th International Conference on Advanced Robotics and Mechatronics (ICARM), Toyonaka, Japan, 3–5 July 2019; pp. 118–123.
44. Xie, H.; Li, G.; Zhao, X.; Li, F. Prediction of Limb Joint Angles Based on Multi-Source Signals by GS-GRNN for Exoskeleton Wearer. *Sensors* **2020**, *20*, 1104. [CrossRef]
45. Rabe, K.G.; Fey, N.P. Evaluating Electromyography and Sonomyography Sensor Fusion to Estimate Lower-Limb Kinematics Using Gaussian Process Regression. *Front. Robot. AI.* **2022**. [CrossRef] [PubMed]
46. Fougner, A.; Scheme, E.; Chan, A.D.; Englehart, K.; Stavdahl, X. Resolving the Limb Position Effect in Myoelectric Pattern Recognition. *IEEE Trans. Neural Syst. Rehabil. Eng.* **2011**, *19*, 644–651. [CrossRef] [PubMed]
47. Guo, B.; Mao, Y.; Han, J.; Li, X.; Ma, J. Human-robot interactive information sensing system for gait rehabilitation training robot. *Chin. J. Eng. Des.* **2019**, *26*, 252–259.
48. Gong, C.; Xu, D.; Zhou, Z.; Nicola, V.; Wang, Q. BPNN-Based Real-Time Recognition of Locomotion Modes for an Active Pelvis Orthosis with Different Assistive Strategies. *Int. J. Hum. Robot.* **2020**, *17*, 2050004. [CrossRef]
49. Zhu, L.; Wang, Z.; Ning, Z.; Zhang, Y.; Liu, Y.; Cao, W.; Wu, X.; Chen, C. A Novel Motion Intention Recognition Approach for Soft Exoskeleton via IMU. *Electronics* **2020**, *9*, 2176. [CrossRef]
50. Pinheiro, C.; Figueiredo, J.; Magalhães, N.; Santos, C.P. Wearable Biofeedback Improves Human-Robot Compliance during Ankle-Foot Exoskeleton-Assisted Gait Training: A Pre-Post Controlled Study in Healthy Participants. *Sensors* **2020**, *20*, 5876. [CrossRef]
51. Xu, W.; Huang, J.; Yan, Q.; Wang, Y.; Tao, C. Flexible and Safe Robot Movement Control Research. *J. Autom.* **2016**, *42*, 1859–1873.
52. Mora-Tola, E.; Loja-Duchi, J.; Ordoñez-Torres, A.; Vázquez-Rodas, A.; Astudillo-Salinas, F.; Minchala, L.I. Robotic Knee Exoskeleton Prototype to Assist Patients in Gait Rehabilitation. *IEEE Lat. Am. Trans.* **2020**, *18*, 1503–1510. [CrossRef]
53. Wu, Q.; Wang, X.; Chen, B.; Wu, H. Development of a Minimal-Intervention-Based Admittance Control Strategy for Upper Extremity Rehabilitation Exoskeleton. *IEEE Trans. Syst. Man Cybern Syst.* **2018**, *48*, 1005–1016. [CrossRef]
54. Duncan, P.W.; Martha, P.; Nelson, S.G. Reliability of the Fugl-Meyer assessment of sensorimotor recovery following cerebrovascular accident. *Phys. Therapy* **1983**, *10*, 1606–1610. [CrossRef]
55. Salarian, A.; Russmann, H.; Vingerhoets, F.; Dehollain, C.; Blanc, Y.; Burkhard, P.; Aminian, K. Gait assessment in Parkinson's disease: Toward an ambulatory system for long-term monitoring. *IEEE Trans. Biomed. Eng.* **2004**, *51*, 1434–1443. [CrossRef]
56. Anaya-Reyes, F.; Cheng, H.; Thangavel, P.; Yu, H. The Shared Effects of Active Body Weight Support and Robot-Applied Resistance/Assistance on Temporal Gait Parameters and Gait Related Muscle Activity. In Proceedings of the 2018 3rd International Conference on Advanced Robotics and Mechatronics (ICARM), Singapore, 8–20 July 2018; pp. 386–391.
57. Alberto, R.; Martina, F.; Tommaso, S.; Susanna, S.; Silvia, M.; Adele, D.; Michela, C.; Maurizio, P.; Enrico, C.; Enrico, B.; et al. Evaluation of gait in Duchenne Muscular Dystrophy: Relation of 3D gait analysis to clinical assessment. *Neuromuscul. Disord.* **2019**, *29*, 920–929.
58. Ma, Y.; Wu, X.; Wang, C.; Yi, Z.; Liang, G. Gait Phase Classification and Assist Torque Prediction for a Lower Limb Exoskeleton System Using Kernel Recursive Least-Squares Method. *Sensors* **2019**, *19*, 5449. [CrossRef]
59. Chomiak, T.; Sidhu, A.S.; Watts, A.; Su, L.; Graham, B.; Wu, J.; Classen, S.; Falter, B.; Hu, B. Development and Validation of Ambulosono: A Wearable Sensor for Bio-Feedback Rehabilitation Training. *Sensors* **2019**, *19*, 686. [CrossRef]
60. Tran, V.-T.; Sasaki, K.; Yamamoto, S.-I. Influence of Body Weight Support Systems on the Abnormal Gait Kinematic. *Appl. Sci.* **2020**, *10*, 4685. [CrossRef]
61. Park, C.; Oh-Park, M.; Bialek, A. Abnormal synergistic gait mitigation in acute stroke using an innovative ankle–knee–hip interlimb humanoid robot: A preliminary randomized controlled trial. *Sci. Rep.* **2021**, *11*, 22823. [CrossRef]
62. Sconza, C.; Negrini, F.; Di Matteo, B.; Borboni, A.; Boccia, G.; Petrikonis, I.; Stankevičius, E.; Casale, R. Robot-Assisted Gait Training in Patients with Multiple Sclerosis: A Randomized Controlled Crossover Trial. *Medicina* **2021**, *57*, 713. [CrossRef]

63. Choi, M.; Ahn, N.; Park, J.; Kim, K. 12-Week Exercise Training of Knee Joint and Squat Movement Improves Gait Ability in Older Women. *Int. J. Environ. Res. Public Health* **2021**, *18*, 1515. [CrossRef]
64. Yeon-Gyo, N.; Jin, W.L.; Jin, W.P.; Ho, J.L.; Ki, Y.N.; Jun, H.P.; Chang, S.Y.; Myong, R.C.; Bum, S.K. Effects of Electromechanical Exoskeleton-Assisted Gait Training on Walking Ability of Stroke Patients: A Randomized Controlled Trial. *Arch. Phys. Med. Rehabil.* **2019**, *100*, 26–31.
65. Wang, Y.; Mukaino, M.; Hirano, S.; Tanikawa, H.; Yamada, J.; Ohtsuka, K.; Ii, T.; Saitoh, E.; Otaka, Y. Persistent Effect of Gait Exercise Assist Robot Training on Gait Ability and Lower Limb Function of Patients With Subacute Stroke: A Matched Case–Control Study With Three-Dimensional Gait Analysis. *Front. Neurorobot.* **2020**, *14*, 42. [CrossRef]
66. Tamburella, F.; Tagliamonte, N.L. Neuromuscular Controller Embedded in a Powered Ankle Exoskeleton: Effects on Gait, Clinical Features and Subjective Perspective of Incomplete Spinal Cord Injured Subjects. *IEEE Trans. Neural Syst. Rehabil. Eng.* **2020**, *28*, 1157–1167. [CrossRef]
67. Wahid, F.; Begg, R.K.; Hass, C.J.; Halgamuge, S.; Ackland, D.C. Classification of Parkinson's Disease Gait Using Spatial-Temporal Gait Features. *IEEE J. Biomed. Health Inform.* **2015**, *19*, 1794. [CrossRef]
68. Rehman, R.Z.U.; Din, S.D.; Guan, Y.; Yarnall, A.J.; Shi, J.Q.; Rochester, L. Selecting Clinically Relevant Gait Characteristics for Classification of Early Parkinson's Disease: A Comprehensive Machine Learning Approach. *Sci. Rep.* **2019**, *9*, 17269. [CrossRef]
69. Herfurth, M.; Godau, J.; Kattner, B.; Rombach, S.; Grau, S.; Maetzler, W.; Berg, D. Gait velocity and step length at baseline predict outcome of Nordic walking training in patients with Parkinson's disease. *Parkinsonism Relat. Disord.* **2015**, *21*, 413–416. [CrossRef]
70. Carlotta, C.; Diego, T.; Maurizio, S.; Adriana, M.; Jose, G.; Francisco, G.; Jose, P. IMU-Based Classification of Parkinson's Disease From Gait: A Sensitivity Analysis on Sensor Location and Feature Selection. *IEEE J. Biomed. Health Inform.* **2018**, *22*, 1765–1774.
71. Wang, L.; Sun, Y.; Li, Q.; Liu, T.; Yi, J. Two Shank-Mounted IMUs-Based Gait Analysis and Classification for Neurological Disease Patients. *IEEE Robot. Autom. Lett.* **2020**, *5*, 1976–1979. [CrossRef]
72. Skvortsov, D.V.; Kaurkin, S.N.; Ivanova, G.E. A Study of Biofeedback Gait Training in Cerebral Stroke Patients in the Early Recovery Phase with Stance Phase as Target Parameter. *Sensors* **2021**, *21*, 7217. [CrossRef]
73. Saito, H.; Yokoyama, H.; Sasaki, A.; Kato, T.; Nakazawa, K. Flexible Recruitments of Fundamental Muscle Synergies in the Trunk and Lower Limbs for Highly Variable Movements and Postures. *Sensors* **2021**, *21*, 6186. [CrossRef] [PubMed]
74. Cheung Vincent, C.K.; Turolla, A.; Agostini, M.; Silvoni, S.; Bennis, C.; Kasi, P.; Paganoni, S.; Bonato, P.; Bizzi, E. Muscle synergy patterns as physiological markers of motor cortical damage. *Proc. Natl. Acad. Sci. USA* **2012**, *109*, 14652–14656. [CrossRef] [PubMed]
75. Safavynia, S.; Torres-Oviedo, G.; Ting, L. Muscle Synergies: Implications for Clinical Evaluation and Rehabilitation of Movement. *Top. Spinal Cord Inj. Rehabil.* **2011**, *17*, 16–24. [CrossRef]
76. Rinaldi, L.; Yeung, L.F.; Lam, C.H.; Pang, M.; Cheung, V. Adapting to the Mechanical Properties and Active Force of an Exoskeleton by Altering Muscle Synergies in Chronic Stroke Survivors. *IEEE Trans. Neural Syst. Rehabil. Eng.* **2020**, *28*, 2203–2213. [CrossRef]
77. Longatelli, V.; Pedrocchi, A.; Guanziroli, E. Robotic Exoskeleton Gait Training in Stroke: An Electromyography-Based Evaluation. *Front. Neurorobot.* **2021**, *15*, 733738. [CrossRef]
78. Seo, J.-W.; Kim, H.-S. Biomechanical Analysis in Five Bar Linkage Prototype Machine of Gait Training and Rehabilitation by IMU Sensor and Electromyography. *Sensors* **2021**, *21*, 1726. [CrossRef]
79. Li, T.; Li, Q.; Liu, T. An actuated dissipative spring-mass walking model: Predicting human-like ground reaction forces and the effects of model parameters. *J. Biomech.* **2019**, *90*, 58–64. [CrossRef]
80. Smyrli, A.; Papadopoulos, E. A methodology for the incorporation of arbitrarily-shaped feet in passive bipedal walking dynamics. In Proceedings of the 2020 IEEE International Conference on Robotics and Automation (ICRA), Paris, France, 31 May–31 August 2020; pp. 8719–8725.
81. Cifuentes, C.A.; Rodriguez, C.; Frizera-Neto, A.; Bastos, T.F.; Carelli, R. Multimodal Human–Robot Interaction for Walker-Assisted Gait. *IEEE Syst. J.* **2017**, *10*, 933–943. [CrossRef]
82. Kolaghassi, R.; Al-Hares, M.K.; Sirlantzis, K. Systematic Review of Intelligent Algorithms in Gait Analysis and Prediction for Lower Limb Robotic Systems. *IEEE Access* **2021**, *9*, 113788–113812. [CrossRef]
83. Hu, X.; Zeng, X.; Xu, Y. A Soft Robotic Intervention for Gait Enhancement in Older Adults. *IEEE Trans. Neural Syst. Rehabil. Eng.* **2021**, *29*, 1838–1847. [CrossRef]
84. Yu, H.; Spenko, M.; Dubowsky, S. An Adaptive Shared Control System for an Intelligent Mobility Aid for the Elderly. *Auton. Robot.* **2003**, *15*, 53–66. [CrossRef]
85. Tao, W.; Zhang, J.; Li, G.; Liu, T.; Liu, F.; Yi, J. A wearable sensor system for lower-limb rehabilitation evaluation using the GRF and CoP distributions. *Meas. Sci. Technol.* **2015**, *27*, 025701. [CrossRef]
86. Ye, J.; Chen, G.; Liu, Q. An Adaptive Shared Control of a Novel Robotic Walker for Gait Rehabilitation of Stroke Patients. In Proceedings of the 2018 IEEE International Conference on Intelligence and Safety for Robotics (ISR), Shenyang, China, 24–27 August 2018; pp. 373–378.
87. Miyake, T.; Kobayashi, Y.; Fujie, M.G.; Sugano, S. Intermittent Force Application of Wire-Driven Gait Training Robot to Encourage User to Learn an Induced Gait. In Proceedings of the 2018 IEEE International Conference on Robotics and Biomimetics (ROBIO), Kuala Lumpur, Malaysia, 12–15 December 2018; pp. 433–438.

88. Zhao, D.; Yang, J.; Okoye, M.O.; Wang, S. Walking Assist Robot: A Novel Non-Contact Abnormal Gait Recognition Approach Based on Extended Set Membership Filter. *IEEE Access* **2019**, *7*, 76741–76753. [CrossRef]
89. Mu, Z.; Fang, J.; Zhang, Q. Admittance Control of the Ankle Mechanism in a Rotational Orthosis for Walking with Arm Swing. In Proceedings of the 2019 IEEE 16th International Conference on Rehabilitation Robotics (ICORR), Toronto, ON, Canada, 24–28 June 2019; pp. 709–714.
90. Guo, B.; Han, J.; Li, X.; Yan, L. Human-robot interactive control based on reinforcement learning for gait rehabilitation training robot. *Int. J. Adv. Robot. Syst.* **2019**, *16*, 1729881419839584.
91. Kim, J.Y.; Kim, J.J.; Park, K. Gait Training Algorithm of an End-Effector Typed Hybrid Walking Rehabilitation Robot. *Int. J. Precis. Eng. Manuf.* **2019**, *20*, 1767–1775. [CrossRef]
92. Mu, Z.; Zhang, Q.; Yang, G.; Xie, L.; Fang, J. Development of an Improved Rotational Orthosis for Walking With Arm Swing and Active Ankle Control. *Front. Neurorobot.* **2020**, *14*, 17. [CrossRef]
93. Shunki, I.; Jun, N.; Yasuhisa, H.; Toshio, F.; Masanori, T.; Izumi, K. Admittance control based robotic clinical gait training with physiological cost evaluation. *Robot. Auton. Syst.* **2020**, *123*, 103326.
94. Zhang, Y.; Li, S.; Nolan, K.J.; Zanotto, D. Reinforcement Learning Assist-as-needed Control for Robot Assisted Gait Training. In Proceedings of the 2020 8th IEEE RAS/EMBS International Conference for Biomedical Robotics and Biomechatronics (BioRob), New York, NY, USA, 29 November–1 December 2020; pp. 785–790.
95. Zhong, B.; Cao, J.; Guo, K.; Andrew, M.; Peng, Y.; Miao, Q.; Xie, S.; Zhang, M. Fuzzy logic compliance adaptation for an assist-as-needed controller on the Gait Rehabilitation Exoskeleton (GAREX). *Robot. Auton. Syst.* **2020**, *133*, 103642. [CrossRef]
96. Hidayah, R.; Bishop, L.; Jin, X.; Chamarthy, S.; Stein, J.; Agrawal, S.K. Gait Adaptation Using a Cable-Driven Active Leg Exoskeleton (C-ALEX) With Post-Stroke Participants. *IEEE Trans. Neural Syst. Rehabil. Eng.* **2020**, *28*, 1984–1993. [CrossRef]
97. Lopes, J.; Pinheiro, C.; Figueiredo, J.; Reis, L.P.; Santos, C.P. Assist-as-needed Impedance Control Strategy for a Wearable Ankle Robotic Orthosis. In Proceedings of the 2020 IEEE International Conference on Autonomous Robot Systems and Competitions (ICARSC), Ponta Delgada, Portugal, 15–17 April 2020; pp. 10–15.
98. Zhang, H.; Yin, Y.; Chen, Z.; Zhang, Y.; Rao, A.K.; Guo, Y.; Zanotto, D. Wearable Biofeedback System to Induce Desired Walking Speed in Overground Gait Training. *Sensors* **2020**, *20*, 4002. [CrossRef]
99. Scheidig, A.; Schütz, B.; Trinh, T.Q.; Vorndran, A.; Mayfarth, A.; Sternitzke, C.; Röhner, E.; Gross, H.-M. Robot-Assisted Gait Self-Training: Assessing the Level Achieved. *Sensors* **2021**, *21*, 6213. [CrossRef] [PubMed]
100. Ma, Y.; Wu, X. Online Gait Planning of Lower-Limb Exoskeleton Robot for Paraplegic Rehabilitation Considering Weight Transfer Process. *IEEE Trans. Autom. Sci. Eng.* **2021**, *18*, 414–425. [CrossRef]
101. Chen, Z.; Guo, Q.; Xiong, H. Control and Implementation of 2-DOF Lower Limb Exoskeleton Experiment Platform. *Chin. J. Mech. Eng.* **2021**, *34*, 22. [CrossRef]
102. Mayag, L.J.A.; Múnera, M.; Cifuentes, C.A. Human-in-the-Loop Control for AGoRA Unilateral Lower-Limb Exoskeleton. *J. Intell Robot Syst.* **2022**, *104*, 3. [CrossRef]
103. Inoue, J.; Kimura, R. Development of a Gait Rehabilitation Robot Using an Exoskeleton and Functional Electrical Stimulation: Validation in a Pseudo-paraplegic Model. *Prog. Rehabil. Med.* **2022**, *7*, 20220001. [CrossRef]
104. Gianluca, L.; Caroline, P.B. Pelvic floor muscle training and postural balance in elderly women: An exploratory single arm trial. *J. Bodyw. Mov. Ther.* **2022**, *29*, 279–285.
105. Nevisipour, M.; Honeycutt, C.F. Investigating the underlying biomechanical mechanisms leading to falls in long-term ankle-foot orthosis and functional electrical stimulator users with chronic stroke. *Gait Posture* **2022**, *92*, 144–152. [CrossRef]
106. Riener, R.; Lünenburger, L.; Maier, I.C.; Colombo, G.; Dietz, V. Locomotor Training in Subjects with Sensori-Motor Deficits: An Overview of the Robotic Gait Orthosis Lokomat. *J. Healthc. Eng.* **2010**, *1*, 517674. [CrossRef]
107. Loredana, R.; Roberto, P.; Flavia, O.; Alfredo, M.; Francesco, C.; Rocco, S.C. A multidisciplinary advanced approach in central pontine myelinolysis recovery: Considerations about a case report. *Disabil. Rehabil. Assist. Technol.* **2020**. [CrossRef]
108. Portaro, S.; Ciatto, L.; Raciti, L.; Aliberti, E.; Aliberti, R.; Naro, A.; Calabrò, R.S. A Case Report on Robot-Aided Gait Training in Primary Lateral Sclerosis Rehabilitation: Rationale, Feasibility and Potential Effectiveness of a Novel Rehabilitation Approach. *Innov Clin. Neurosci.* **2021**, *18*, 15–19.
109. Naro, A.; Billeri, L.; Manuli, A. Breaking the ice to improve motor outcomes in patients with chronic stroke: A retrospective clinical study on neuromodulation plus robotics. *Neurol Sci.* **2021**, *42*, 2785–2793. [CrossRef]
110. Alshamaa, D.; Soubra, R.; Chkeir, A. A Radar Sensor for Automatic Gait Speed Analysis in Walking Tests. *IEEE Sens. J.* **2021**, *21*, 13886–13894. [CrossRef]
111. Di Nardo, F.; Cucchiarelli, A.; Scalise, L.; Morbidoni, C. Measurement of Stride Time by Machine Learning: Sensitivity Analysis for the Simplification of the Experimental Protocol. *IEEE Trans. Instrum. Meas.* **2022**, *71*, 1–9. [CrossRef]

electronics

MDPI

Article

Human Joint Torque Estimation Based on Mechanomyography for Upper Extremity Exosuit

Yongjun Shi [1], Wei Dong [1], Weiqi Lin [1], Long He [2], Xinrui Wang [2], Pengjie Li [2] and Yongzhuo Gao [1,*]

1 State Key Laboratory of Robotics and System, Harbin Institute of Technology (HIT), Harbin 150001, China; 19b908038@stu.hit.edu.cn (Y.S.); dongwei@hit.edu.cn (W.D.); 22b308009@stu.hit.edu.cn (W.L.)

2 Weapon Equipment Research Institute, China South Industries Group Corporation, Beijing 102202, China; helong208@126.com (L.H.); xinruii@163.com (X.W.); lpj208@163.com (P.L.)

* Correspondence: gaoyongzhuo@hit.edu.cn

Abstract: Human intention recognition belongs to the algorithm basis for exoskeleton robots to generate synergic movements and provide corresponding assistance. In this article, we acquire and analyze the mechanomyography (MMG) to estimate the current joint torque and apply this method to the rehabilitation training research of the upper extremity exosuit. In order to obtain relatively pure biological signals, a MMG processing method based on the Hilbert-Huang Transform (HHT) is proposed to eliminate the mixed noise and motion artifacts. After extracting features and forming the dataset, a random forest regression (RFR) model is designed to build the mapping relationship between MMG and human joint output through offline learning. In addition, an upper extremity exosuit is constructed for multi-joint assistance. Based on the above research, we develop a torque estimation-based control strategy and make it responsible for the intention understanding and motion servo of this customized system. Finally, an actual test verifies the accuracy and reliability of this recognition algorithm, and an efficiency evaluation experiment also proves the feasibility for power assistance.

Keywords: joint torque estimation; upper extremity exosuit; mechanomyography signal processing; rehabilitation training; MMG; human movement assistance

Citation: Shi, Y.; Dong, W.; Lin, W.; He, L.; Wang, X.; Li, P.; Gao, Y. Human Joint Torque Estimation Based on Mechanomyography for Upper Extremity Exosuit. *Electronics* **2022**, *11*, 1335. https://doi.org/10.3390/electronics11091335

Academic Editors: João Paulo Morais Ferreira; Tao Liu and Nicola Francesco Lopomo

Received: 25 February 2022
Accepted: 20 April 2022
Published: 22 April 2022

1. Introduction

As is the case for most of the potential technical equipment for rehabilitation training and movement assistance, the exoskeleton system has always attracted attention in related research fields for the elderly and the disabled [1,2]; however, with some key issues not being resolved, the performance of this wearable robot remains relatively limited [3]. Among them, human intention perception belongs to a pretty critical research point and needs to be studied further. Traditional methods try to achieve this function by monitoring limb kinematics data [4] or human-machine interaction information [5], but problems such as response lag greatly restrict its actual effect. In recent years, recognition methods based on biological signals have emerged and gained wide attention [6], which are expected to realize more effective intention understanding.

Biological signals are often generated before the execution of corresponding actions, which show a certain degree of motion predictability and can make intention recognition more timely and accurate. In addition to eye tracking and galvanic skin response (GSR) [7], which are not suitable for combining with exoskeleton control technology, commonly used types in current research mainly contain electroencephalogram (EEG) [8], surface electromyogram (sEMG) [9], and mechanomyography [10]. The EEG signal originates from the potential of the external electrical field that fluctuates around nerve cells in the brain, and is often applied to classify specific movement patterns [11,12]. Due to its instability and susceptibility to interference, EEG-based methods for data collection, signal processing, and intent identification needs to be further improved and optimized. The sEMG signal represents the total action potentials of different motor units innervated by certain motor neurons, which can be obtained through electrodes placed on the corresponding muscle tissue [13]. It has the advantage of high sensitivity and low delay, but many factors will diminish its effectiveness for data collection, such as skin surface cleanliness, air humidity, and patch electrode position [14]. MMG reflects the mechanical vibration signal generated when the skeletal muscle contracts, and it contains abundant information, such as the number of muscle fibers participating in the exercise, and the amplitude and frequency of vibration [15]. Compared with the above two signals, it has the following advantages [16].

- The measuring device can be used directly without touching the skin, which simplifies the preparation before data collection;
- the signal has strong anti-interference ability and will not be affected by environmental variables such as sweat, humidity, or electromagnetics;
- the equipment cost is pretty low because the data collection task can be completed by using an acceleration sensor that meets the accuracy requirements.

However, there are still some limitations in the practical application of MMG. For example, it is easily contaminated by low-frequency motion artifacts, and sensitive to sudden step noises; therefore, a suitable signal processing algorithm is needed to extract a relatively pure sequence.

At present, classification and regression algorithms have been widely used in many studies, such as the trajectory control of a redundant robot [17] and hand gesture recognition for teleoperated surgical robot systems [18], which can also complete the analysis of human motion intention. As a relatively common and easy-to-use method, motion pattern recognition [19,20] aims at establishing the mapping relationship from sensor information to finite human motion states through classification models. Then, the control subsystem will generate corresponding commands according to current motion pattern and will deliver them to underlying drivers. Joint angle prediction attempts to calculate the limb position at the next moment based on regression models [21,22], which can effectively avoid the response lag of the exosuit. After that, the motion state of the power-assisted system can be dynamically adjusted through position closed-loop control. Joint torque estimation also belongs to a direct and effective method of obtaining intentions [23,24]. The desired torque can be calculated through the Hill type model [25] or obtained from the biological signals using machine learning algorithms [26], and can act as the input parameter of torque closed-loop control to regulate the motor output. The last method will provide a reference for the direct control of external torque that assists joint motion, which is quite practical for upper extremity exosuits that need to achieve an expected power-assisted efficiency.

If we want to use the abovementioned MMG signal to estimate the joint torque, it is obviously quite difficult through conventional mathematical derivation. The machine learning algorithm can train the mapping model very well based on the existing data, and it shows an excellent fitting ability in many research fields, such as breathing pattern detection [27] and human activity identification [28]; therefore, this method should be able to describe the complex and nonlinear relationship between MMG and joint torques.

The inherent characteristics of the soft exosuit based on Bowden cables greatly increases the difficulty for designing control strategies [29,30]. The gravity compensation algorithm is a simple and common method for motion generation, which will output an active torque to offset the joint load imposed by the limb weight [31,32]; however, due to the large error of compensation model and the ignorance of dynamic characteristics, it may cause the system response to, more or less, mismatch the joint movement. The exosuit named CRUX [33] tries to control the target arm to follow the reference trajectory of the healthy one, consequently completing the active rehabilitation training process [34]. This method will limit the subjective initiative of the wearer to a certain extent, and is not suitable for situations where both arms need assistance. Some scholars from Italy have proposed a threshold method based on sEMG [35]. When the signal amplitude of the wearer's measured muscle exceeds the set value, the wearable system starts to produce a power-assisted effect for motion immediately. The most obvious disadvantage of this method is that it only outputs a constant driving force, but the required joint torque changes dynamically under different motion states. In general, the control logic for the upper extremity exosuit still has some defects, and remains to be further explored.

In this paper, we intend to complete data preprocessing and feature extraction using MMG, then establish the mapping relationship from this signal to the joint torque based on the machine learning algorithm, and finally apply it to the rehabilitation training research of upper extremity exosuit. MMG is collected through an inertial measurement unit (IMU) and synthesized by the linear accelerations along three axes. The HHT will filter this original signal to obtain a relatively pure result. We extract three features from the processed data and combine them with the collected joint torque to form data sets for training and testing. A RFR model is designed as the algorithm framework for joint torque estimation, and its parameters are determined through offline training. According to the above research results, a control algorithm based on joint torque estimation will take charge of the motion control for an upper extremity exosuit. Eventually, some experiments are carried out to test the accuracy of torque estimation and the efficiency of power assistance. The main contributions and highlights of this study are summarized as follows.

- We have attempted to use the MMG signal as the medium for the exosuit to understand human intentions, and to demonstrate the feasibility and effectiveness of this approach.
- A series of MMG-related methods for signal acquisition, filtering, and feature extraction have been developed.
- A regression model reflecting the nonlinear relationship between muscle activation and joint output is constructed.
- A torque estimation-based control algorithm is designed and applied to the multi-joint motion assistance of upper extremity exosuit, which can significantly amplify the limb strength.

The remaining research content of this article is organized as follows. Section 2 describes the measurement and calculation methods for MMG, its corresponding joint torque, and how to process the original signal to construct data sets. Section 3 introduces the design details of RFR model and uses a large amount of test data to fit the desired quantitative relationship. In Section 4, we have developed a control strategy for the upper extremity exosuit based on torque estimation. Section 5 proves the feasibility of the above methods through some experiments. Section 6 is the conclusion.

2. Data Sets Acquisition

2.1. Raw Information Collection

In order to obtain the MMG signal and corresponding joint torque at the same time, we built a measurement platform for joint information collection. Figure 1a shows how to use this device to get relevant data about elbow static flexion and extension.

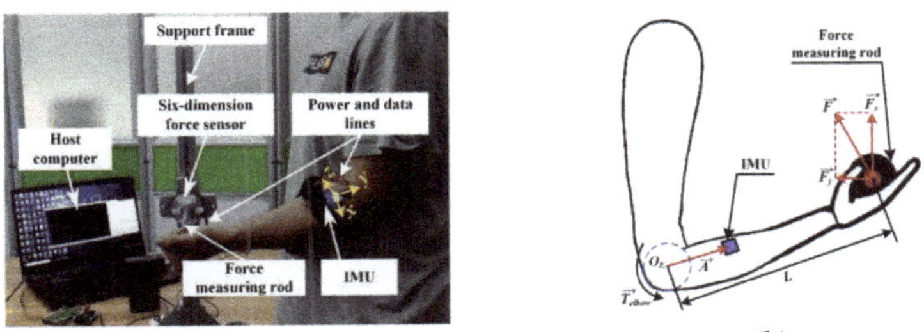

(a) (b)

Figure 1. Information collection process during elbow static flexion/extension: (**a**) measurement platform; (**b**) joint torque calculation model for elbow.

During this process, a six-dimension force sensor and a high-performance IMU are responsible for detecting forces along three directions at the end of limb, and gathering MMG at the brachioradialis of forearm, respectively. When subjects hold the force measuring rod and try to perform specific actions, this platform will transmit corresponding data to the host computer through a data line and save them in files.

With readings from a six-dimension force sensor (F_x and F_y) as input, the joint torque ($\vec{T}_{elbow}$) can be calculated by a certain mathematical relation. The mechanical model of the elbow is shown in Figure 1b, where $\vec{A}$, $\vec{F}$, and L, respectively, represent the current posture vector of human arm, the resultant force vector at the end, and the length of forearm. Then, we can dynamically obtain the joint torque values during elbow static flexion and extension through the following formulas.

$$\vec{F} = \vec{F}_x + \vec{F}_y \tag{1}$$

$$\vec{T}_{elbow} = L\left(\frac{\vec{A}}{|\vec{A}|} \times \vec{F}\right) \tag{2}$$

During dynamic flexion/extension without using the platform, it can be seen in Figure 2 that when the measured muscle contracts or relaxes and drives the human joint to rotate, the IMU can perceive some regular acceleration signals in the x, y, and z directions synchronously, and the same is true for the process of static joint output. In order to integrate all the effective information, we calculate the sum of linear acceleration vectors along three axes (ACC_x, ACC_y, and ACC_z), and take its modulus as the original MMG signal ($MMG(t)$), which can be expressed as the following equation.

$$MMG(t) = \sqrt{ACC_x^2 + ACC_y^2 + ACC_z^2} \tag{3}$$

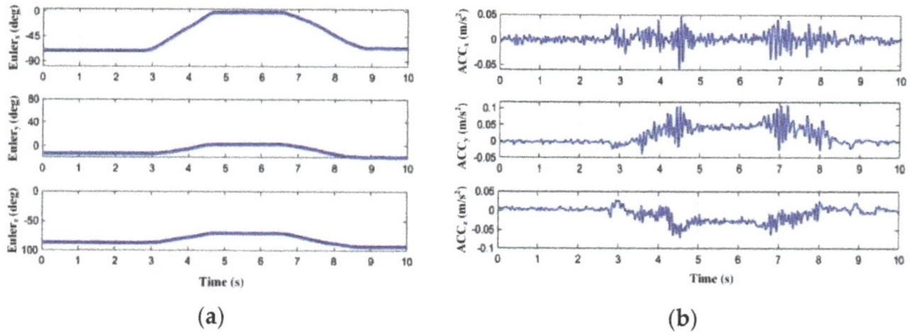

Figure 2. The IMU data during elbow dynamic flexion/extension with low strength: (**a**) three-axis euler angles; (**b**) three-axis accelerations.

2.2. MMG Signal Processing

However, the method of using linear acceleration values for MMG characterization inevitably mixes low-frequency artifacts into the collected original signal, such as gravitational acceleration, IMU posture changing artifacts caused by muscle deformation, and motion artifacts introduced by upper limb movements. Moreover, high-frequency white noise may also be superimposed on it. In order to improve the quality of data and lay the foundation for subsequent feature extraction, an effective filtering method must be applied to eliminate the abovementioned interferences. Traditional signal processing methods are mostly based on Fourier analysis, but these ones have limited effects in practical applications of processing MMG due to its non-linear and non-stationary characteristics. With reference to related literatures, we decide to use HHT to analyze the original data, which is more suitable for these kinds of signals.

The HHT consists of two parts, namely empirical mode decomposition (EMD) and the Hilbert transform. EMD can decompose a complex signal into a limited number of intrinsic mode functions (IMFs) and a residual based on the local time scale characteristics of itself. The specific implementation steps are as follows. First, we find all the local maximum points and local minimum points of the original MMG signal and fit their respective envelopes through the cubic spline curve. Then, the mean value of the upper and lower envelopes ($U(t)$ and $L(t)$) will be subtracted from the original sequence to get the remaining part with the low frequency information removed ($x(t)$).

$$x(t) = MMG(t) - \frac{1}{2}(U(t) + L(t)) \tag{4}$$

If the number of extreme values and zero crossings on the entire data set of $x(t)$ differs by 0 or 1, and the average value of its two envelopes remains zero at any time, then $IMF_i(t) = x(t)$, otherwise it is necessary to let $MMG(t) = x(t)$, and repeat the above steps until these two conditions are met. Next, we remove the obtained $IMF_i(t)$ from $MMG(t)$ and repeat all the above steps again with the remaining part ($r_i(t)$) to get other IMF components until $r_i(t)$ is a constant or monotonic function. As shown in Figure 3a, the original MMG signal is decomposed into 9 IMFs and 1 residual ($res(t)$) according to the signal frequency, which can be expressed as follows.

$$MMG(t) = \sum_{i=1}^{9} IMF_i(t) + res(t) \tag{5}$$

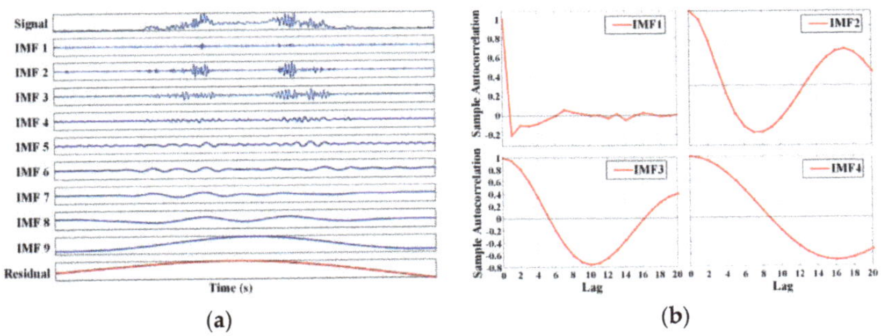

Figure 3. MMG signal processing: (**a**) decomposing result through EMD; (**b**) autocorrelation function curves of the first four IMFs.

After completing the above analysis, it is time to select the IMFs dominated by MMG through certain methods and reorganize them to obtain a relatively pure signal. To eliminate the noise-dominated IMFs, we introduce the concept of autocorrelation ($R_{IMF}(t_1,t_2)$), which reflects the correlation degree of signal values at different times (t_1 and t_2). Its normalized expression form, ($\rho_{IMF}(\tau)$), can be obtained with the following formula, where $\tau = t_1 - t_2$.

$$R_{IMF}(t_1, t_2) = E[IMF(t_1) \cdot IMF(t_2)] \tag{6}$$

$$\rho_{IMF}(\tau) = \frac{R_{IMF}(\tau)}{R_x(0)} \tag{7}$$

If the normalized autocorrelation curve belongs to an impulse function close to the zero point, it can be ascertained that the corresponding IMF is dominated by noise, because noise has randomness and a weak correlation at every moment. After the calculations in Figure 3b, the first IMF can be classified as such a disturbance.

To exclude the IMFs dominated by motion artifacts, we try to find the discrimination basis from the energy distribution of each order component. After calculating the energy value (E^i_{IMF}) of each IMF according to the following Formula (8), it is revealed that energy contained in each IMF has increased significantly, starting from the sixth order. By analyzing the data of different participants, we find that removing these IMFs as motion artifacts can achieve a better result.

In the end, we believe that IMF2~IMF5 can effectively characterize the muscle activity, and the filtered signal can be obtained with them being recombined. It can be seen from Figure 4 that there is a pretty clear correspondence between the processed MMG and joint torque.

$$E^i_{IMF} = \frac{1}{n}\sum\nolimits_{j=1}^{n} [IMF_i(j)]^2 \quad (8)$$

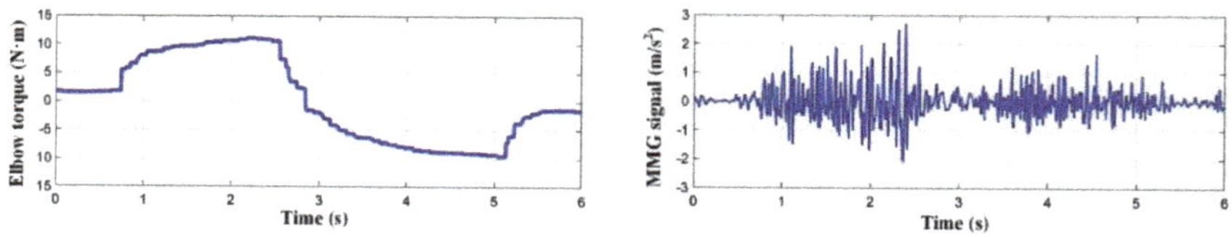

(a) (b)

Figure 4. Comparison between the filtered MMG of the brachioradialis muscle and the corresponding elbow joint torque: (**a**) the change curve of joint torque; (**b**) the change curve of filtered MMG.

2.3. Feature Extraction

Considering that the change in limb strength is often accompanied by the fluctuation of the muscle fiber's vibration amplitude, we select the root mean square (RMS) as the time domain characteristic of MMG. It reflects the effective value of data amplitude and can be calculated with the following formula.

$$RMS_{MMG} = \sqrt{\frac{1}{N}\sum\nolimits_{i=1}^{N} X_i^2} \quad (9)$$

Due to the correlation between muscle activity and its vibration frequency, mean power frequency (MPF) is used to represent the frequency domain characteristic of MMG. It is necessary to perform the Hilbert transform on the filtered MMG to analyze its frequency spectrum, and then integrate it on the time axis to obtain the Hilbert marginal spectrum that characterizes the relationship between signal frequency (f_i) and energy (E_i). Then, MPF can be calculated through the following equation.

$$MPF_{MMG} = \frac{\sum_{i=1}^{N} f_i E_i}{\sum_{i=1}^{N} E_i} \quad (10)$$

The MMG signal may contain information about the number of muscle fibers involved in power output; therefore, we additionally introduce the concept of sample entropy (SampEn) to describe the characteristic from a non-linear perspective, which can quantify the complexity of the time series.

In order to not lose continuity information in the sequence, we apply the sliding window strategy to extract these characteristic values of the filtered MMG signal. The window length and step length are set as 500 ms and 50 ms, respectively. So far, the data set of elbow static flexion/extension is established with RMS, MPF, and SampEn of the MMG signal as features, and joint torque as the label. We can also use similar methods to obtain relevant information of shoulder static flexion/extension and static adduction/abduction.

3. Off-Line Torque Estimation

3.1. Regression Model Design

There is a relatively complicated non-linear relationship between MMG and human joint torque, which seems difficult to accurately describe using the traditional polynomial regression model. We try to introduce a machine learning method to solve this problem, using a large amount of test data to fit the real mapping law. Considering that the joint torque estimation algorithm is oriented to a wearable power-assisted system, the requirement for its stability and reliability must take precedence over that of other aspects. Since RFR (shown as Figure 5) has these advantages, it can be well qualified for this task.

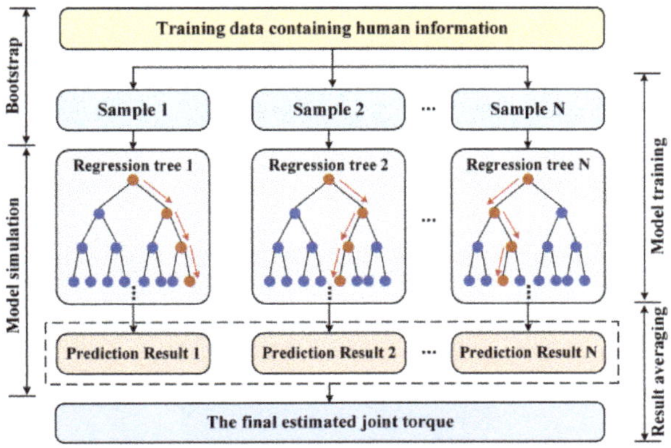

Figure 5. Schematic diagram of RFR algorithm.

The RFR belongs to a bagging type algorithm of ensemble learning, which aims at improving overall performance by packaging and combining several weak models, namely decision trees, into a strong one. The entire model consists of multiple classification and regression trees (CARTs) that are not related to each other. All CARTs jointly determine the final output result. The specific implementation steps of the algorithm are described as follows.

1. Randomly extract any number of samples from the training set to form multiple new sub-training sets;
2. use each sub-training set to train a CART separately. During this process, it is necessary to randomly obtain any number of features from all the features, and then select the optimal segmentation point to cut the subtree;
3. repeat step 2 to obtain multiple trained CARTs;
4. calculate the average of all the CARTs' prediction results and use it as the final estimated value.

Only when more than half of the CARTs make wrong predictions will the output of the RFR model seriously deviate from the true value. Even if an abnormal data point appears, it does not affect the performance of entire algorithm too much, which fully reflects the strong robustness to stop interference signals.

3.2. *Off-Line Training and Testing*

We recruited three healthy adult men to participate in training data acquisition. Based on the abovementioned platform and methods, the information collection experiment for three motion modes obtains 150,000 sets of sample data in total. In accordance with the idea of cross-validation, one-fifteenth of them are selected as the test set, and the remaining data act as the training set. Finally, on the basis of setting the number of sub-regression trees of the RFR to 10, and the minimum leaf size to 1, the training process has been carried out, and the verification result of the test set is also obtained.

Taking shoulder static adduction/abduction as an example, Figure 6a shows that the minimum mean square error (MSE) decreases and tends to be stable with the increase in iterations, and Figure 6b indicates that the difference between the predicted results and the actual values on the test set is relatively small. In general, the model training effect has reached the desired level.

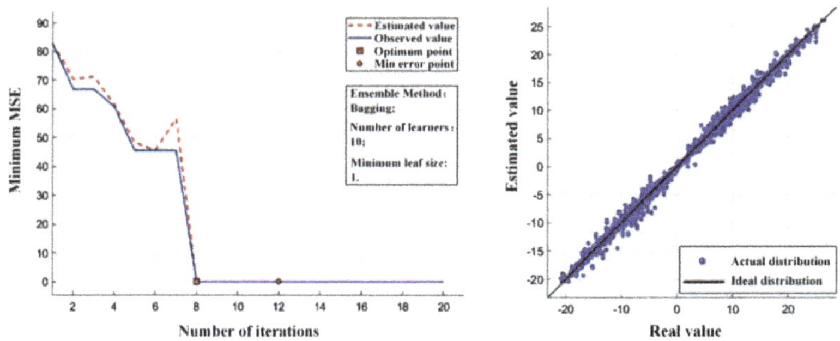

(a) (b)

Figure 6. Machine learning effect of shoulder static adduction/abduction: (**a**) iterative training process with training set; (**b**) verification result of test set.

To measure the predictive performance of trained RFR model, a root mean square error (*RMSE*) and a coefficient of determination (R^2) are introduced as evaluation indexes. The *RMSE* is a commonly used method to express numerical errors, representing the sample standard deviation of the difference between the predicted value and the actual one. It can be calculated by using the following formula.

$$RMSE = \sqrt{\frac{1}{n}\sum\nolimits_{t=1}^{n}(\hat{y}_t - y_t)^2} \tag{11}$$

The R^2 reflects how much the regression relationship can account for changes to the dependent variable. A higher value indicates that the regression model can produce better prediction results. The corresponding calculation process is shown below.

$$R^2 = 1 - \frac{\sum_{t=1}^{n}\hat{y}_t - y_t)^2}{\sum_{t=1}^{n}(y_t - \overline{y})^2} \tag{12}$$

4. Test Platform Construction

4.1. Overview of the Upper Extremity Exosuit

We intend to take advantage of the abovementioned research to design a control logic for an upper extremity exosuit, so that it can perform rehabilitation training functions according to human intentions. As shown in Figure 7, this wearable system aims at providing active assistance for shoulder flexion/extension, shoulder adduction/abduction, and elbow flexion/extension of the left arm.

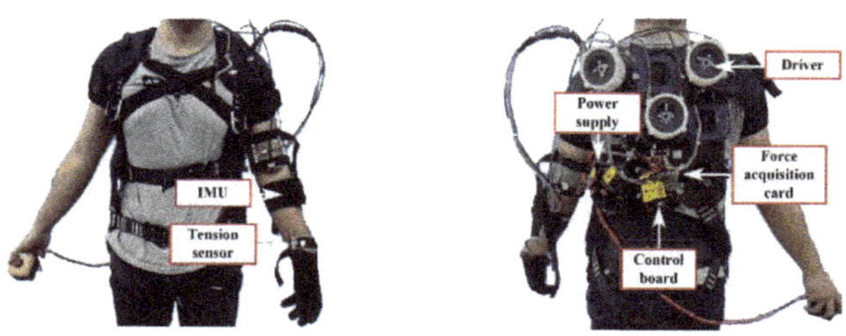

(a) (b)

Figure 7. Pictures of an upper extremity exosuit: (**a**) the front; (**b**) the back.

It contains three sets of cable-driven modules, each of which is responsible for driving the bidirectional motion for one degree of freedom. The sensing network consists of three IMUs, six tension sensors, and three absolute encoders which are integrated in motors, and are in charge of completing multiple tasks, such as MMG signal collection, limb posture perception, human-machine interaction information acquisition, and servo motor state reading. As the main control board, STM32F407IGH6 will serve as the brain of the system to perform core functions such as feature extraction, motion intent identification, and motor servo control. Components communicate with each other through CAN bus for data feedback and instruction delivery.

On the basis of the abovementioned hardware, the exosuit can be driven to assist the human limb coupled with a suitable control algorithm.

4.2. Torque Estimation-Based Control Strategy

As shown in Figure 8, the control logic framework of the upper extremity exosuit consists of two layers, namely, the intent analysis part based on torque estimation, and the motion control part based on torque closed-loop.

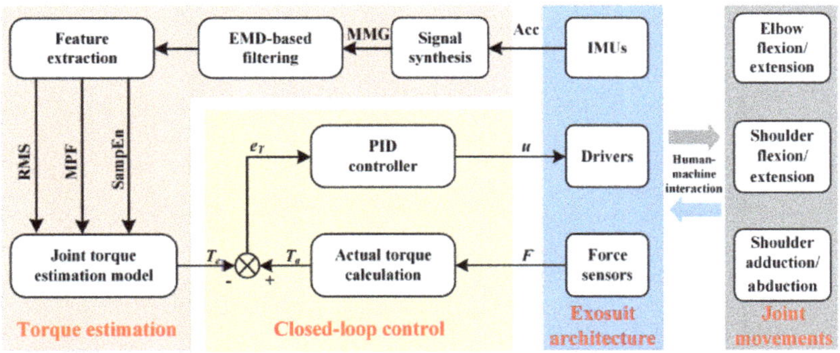

Figure 8. Torque estimation-based control strategy.

When subjects wear this exosuit for collaborative movement, the upper controller will obtain the triaxial accelerations from the target muscle through the accelerometer embedded in the IMU, and synthesize them into an original MMG. After completing the EMD-based filtering operation, it screens out relatively pure signals and extracts the three characteristics including RMS, MPF, and SampEn. The trained RFR model uses these features as an input to estimate the expected joint torque value at the current moment. Finally, the corresponding control commands will be sent to the lower layer.

The lower controller calculates the actual joint torque using the tension sensor readings at the end of the Bowden cable and compares it with the expected joint torque received from the upper layer to obtain their error value. Then, a standard PID algorithm generates motor drive commands based on the torque error, and controls the cable-driven module to provide appropriate assistance at the joint.

To realize this control strategy through a program code, we develop an embedded software based on the μC/OS III operating system. Five sub-tasks, including sensing data reception, signal processing, feature extraction, torque estimation, and motor servo control, are set up in order of priority from high to low. The execution frequency of each is assigned by setting different cycle times. Through the division of the abovementioned sub-task modules, we strengthen the real-time performance of programs under the premise of clarifying the control code logic for the upper extremity exosuit. In addition, it is convenient for subsequent optimization work.

5. Experiment on Exosuit

5.1. Reliability Analysis Experiment for Torque Estimation

The parameters of the RFR model are determined by offline training on a PC. After transplantation to the control system of the exosuit, an evaluation experiment needs to be carried out to examine its actual application effect for different people. We recruited three male volunteers aged 22–27 to complete this experiment. Among them, two subjects (marked as Subject 1 and Subject 2) who have taken part in the training data set collection for torque estimation, are selected to join the experimental group, and one (marked as Subject 3), who did not participate in that process, is assigned to the control group. It is worth noting that volunteers should not have done any high-intensity exercise 24 hours before the tests, to avoid affecting the physiological state of the muscles. During the experiment, they are told to exert an external force that changes approximately in accordance with the sine law on the measurement platform. All subjects knew and agreed with relevant experimental procedures in advance. Research related to this article was approved by the Laboratory Academic Committee of the State Key Laboratory of Robotics and System, Harbin Institute of Technology.

The embedded system, mounted on an upper extremity exosuit, calculates the estimated torque in real time, and sends them to a PC after being processed. The sensor on the measurement platform obtains the force data at the end of the arm, which is converted into the actual torque value in the PC. As it only aims to evaluate the reliability of torque estimation, we have shielded the subtask of the motor servo control in the program, so as to avoid the influence of man-machine coupling.

Figure 9 demonstrates the elbow joint torque estimation results of the upper extremity exosuit on three subjects. In the experimental group, it is obvious that the estimated torque looks very close to the actual value in terms of magnitude and variation trend. Under this condition, the model performance behaves in a relatively stable manner, and the identification result remains rather accurate; however, in the control group, the estimated torque cannot effectively follow the change of the actual value.

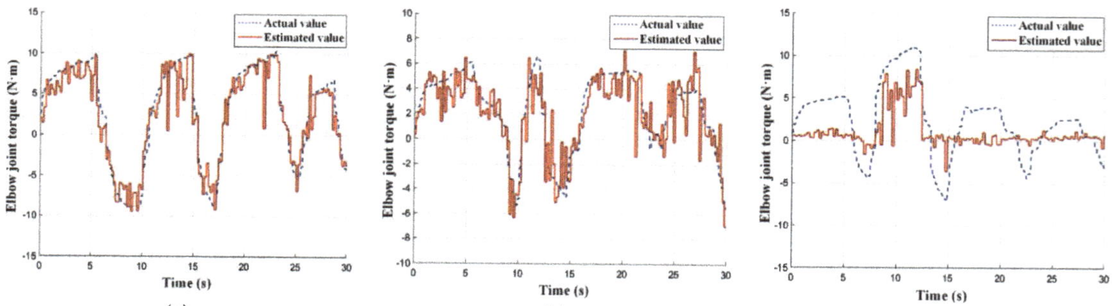

(a) (b) (c)

Figure 9. Elbow joint torque estimation results: (**a**) experimental result of Subject 1; (**b**) experimental result of Subject 2; (**c**) experimental result of Subject 3.

We use the *RMSE* and R^2 introduced above to quantitatively describe the identification effect for different subjects. It can be seen from Table 1 that the *RMSE* of the experimental group is lower than that of the control group, indicating that the error between the actual and estimated value is smaller for the joint torque of Subject 1 and Subject 2. Moreover, the R^2 in the experimental group comes up to 100%, which, when closely compared with the control group, means that the trained RFR model can perform better when utilizing the biological signals of Subject 1 and Subject 2.

Table 1. Evaluation results of torque estimation on three subjects.

Groups	Participants	*RMSE*	R^2
Experimental group	Subject 1	1.9812	0.9532
	Subject 2	1.7008	0.8620
Control group	Subject 3	3.4261	0.6824

From the above qualitative and quantitative description, the following two conclusions can be obtained.

- If we have collected a person's MMG signal for training the RFR model offline, the reliability of the online torque estimation will remain at a pretty high level when they wear the exosuit that uses the trained model;
- utilizing a trained model to estimate the joint torque of unknown subjects online may significantly weaken the effectiveness of identification.

The reason may be that muscle activation varies among different people when they output the same joint torque, or different thicknesses of adipose layers more or less influences MMG propagation; therefore, when using the exosuit for power assistance, it is necessary to independently train a matching torque estimation model for the wearer based on his/her biological information.

5.2. Efficiency Evaluation Experiment for Power Assistance

In order to verify the actual power-assisted effect of this method, we selected a healthy subject, and collected his MMG signals at the brachioradialis, deltoid, and ectopectoralis to customize a set of RFR models for him. After transplanting these trained models to the embedded system, and enabling all the subtasks of control program, the subject wears the exosuit to perform elbow static flexion/extension, shoulder static flexion/extension, and shoulder static adduction/abduction on the measurement platform, and tries to complete three evaluation experiments. Other conditions and requirements are basically the same as the above experiment. An emergency stop switch needs to be held by the right hand all the way through the experiment, to ensure that the experiment can be stopped in time if an accident occurs.

Figure 10 shows the performance evaluation experiments for joint movement assistance. We take three torque values, which are estimated by the RFR model, calculated by the tension sensor on the cable, and converted by the six-dimension force sensor on the measurement platform as human-exerted, exosuit-generated, and the total output, respectively.

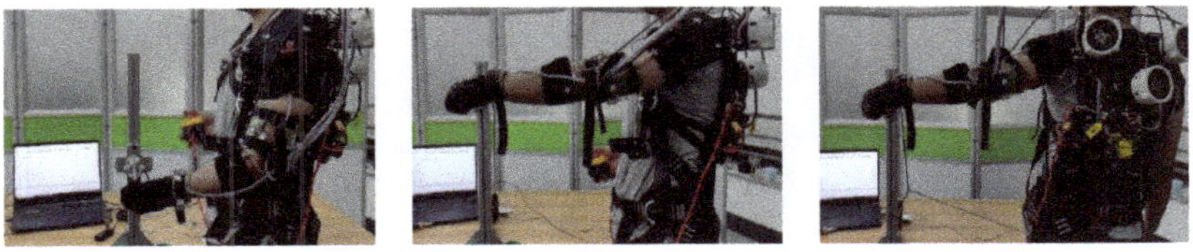

(a) (b) (c)

Figure 10. Actual power-assisted experiments: (**a**) experiment for elbow static flexion/extension; (**b**) experiment for shoulder static flexion/extension; (**c**) experiment for shoulder static adduction/abduction.

Figure 11 describes the changing situation of different torques in the typical time period of each motion mode. Obviously, it can be seen that the upper extremity exosuit can produce additional assistance in the three degrees of freedom of the shoulder and elbow joints, although its actual output is smaller than the torque estimated by the physiological signal. This error can be attributed to the loss of power transmission caused by friction between cable and sheath, or the calculation model deviation induced by suit deformation.

Moreover, the total output far exceeds the human effort, indicating that this exosuit can significantly enhance joint strength.

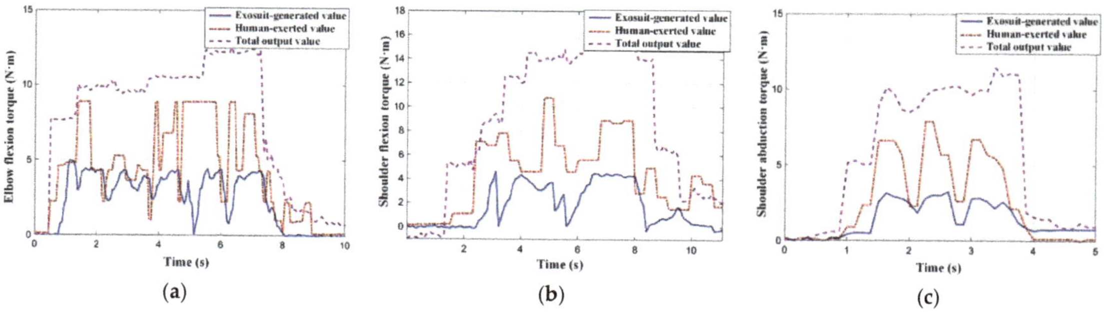

Figure 11. Variations of different torques in each motion mode: (**a**) torque curves for elbow static flexion; (**b**) torque curves for shoulder static flexion; (**c**) torque curves for shoulder static abduction.

We introduce the sum of absolute values (*ASUM*) to quantitatively reflect the average level of the three torques in different motion modes, and the corresponding results are shown in Figure 12.

$$ASUM = \frac{1}{n}\sum_{i=0}^{n}|T_i| \tag{13}$$

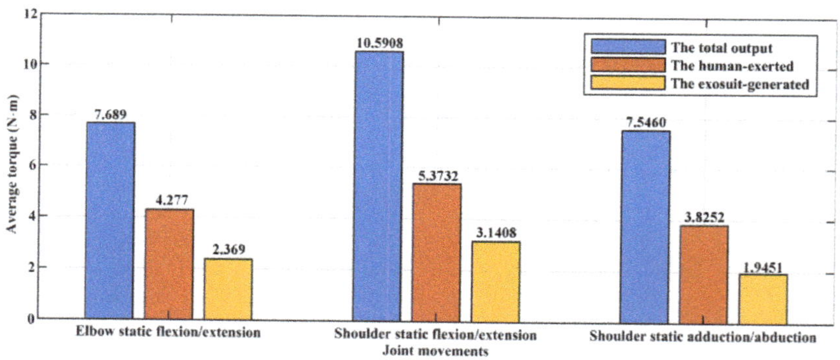

Figure 12. *ASUM* of different torques under each joint movement.

It is obvious that the sum of torque generated by the exosuit, and that exerted by a human, is not equal to the actual total output. The combined effect of factors such as identification error, transmission error, and calculation error, may lead to this gap between the expected and the actual. We expect to describe the power-assisted efficiency (*P*) through analyzing the ratio of $ASUM_{exosuit}$ to $ASUM_{total}$.

$$P = \frac{ASUM_{exosuit}}{ASUM_{total}} \tag{14}$$

The calculation results show that when the upper extremity exosuit independently assists elbow static flexion/extension, shoulder static flexion/extension, and shoulder static adduction/abduction, the corresponding power-assisted efficiencies come up to 30.81%, 29.66%, and 25.78%, respectively. These data mean that when a person is equipped with this wearable robot, the output torque for each joint of the upper limb can be roughly reduced by a quarter to a third.

6. Conclusions and Future Work

In this article, we propose a MMG-based joint torque estimation algorithm which realizes the decoding from biological signals to limb strength, and transplant it into the control system of an exosuit to calculate the motor commands for assisting the multi-joint motion of the upper limb. Two sets of experiments are carried out to test the reliability of torque estimation and the efficiency of power assistance.

The data collection and signal processing methods used in this paper effectively establish the data sets which reflect human body information. The specially designed measurement platform can obtain the MMG of muscles and corresponding joint torque in a relatively accurate and convenient way. The HHT successfully eliminates the interference components in the original MMG signal, and lays a solid foundation for future extraction.

A technical approach to estimate joint torque from the MMG signal is built through training the RFR model offline. The reliability analysis experiment shows that this method can enable the exosuit to accurately identify the wearer's current joint torque, but the model parameters need to be specially trained for everyone.

A torque estimation-based control strategy is successfully applied to the motion control of the upper extremity exosuit. The efficiency evaluation experiment indicates that the exosuit using this algorithm can significantly enhance the limb strength of wearers.

Based on the actual execution of the research, we believe that the current work has the following limitations. First of all, the nonlinear disturbance caused by transmission friction and motion hysteresis significantly reduces the control performance and power-assisted efficiency of the upper extremity exosuit. In addition, the MMG-based torque estimation algorithm has limitations in its application. The model parameters may be trained separately for each person, and even each muscle.

Therefore, future work and research directions should aim to break through the above limitations. First, an error compensation algorithm for this cable-driven system should be introduced into the control logic, in an attempt to offset interferences caused by nonlinear characteristics. Second, more general intention recognition algorithms need to be studied further, which can meet the usage requirements of every wearer without additional training preparation.

Author Contributions: Methodology, data analysis, writing, Y.S.; conceptualization, project administration, W.D.; experimental verification, data preprocessing, W.L.; resources, supervision, L.H.; schematic analysis, resources, X.W.; structure design, resources, P.L.; formal analysis, software, Y.G. All authors have read and agreed to the published version of the manuscript.

Funding: This work was funded by the pre-research project in the manned spacefight field of China (Project Number 020202).

Informed Consent Statement: Informed consent was obtained from all subjects involved in the study.

Acknowledgments: The authors thank all those who have provided suggestions and assistance for this research and paper.

Conflicts of Interest: The authors declare no conflict of interest.

References

1. Majidi Fard Vatan, H.; Nefti-Meziani, S.; Davis, S.; Saffari, Z.; El-Hussieny, H. A review: A Comprehensive Review of Soft and Rigid Wearable Rehabilitation and Assistive Devices with a Focus on the Shoulder Joint. *J. Intell. Robot. Syst.* **2021**, *102*, 9. [CrossRef]
2. Sanjuan, J.D.; Castillo, A.D.; Padilla, M.A.; Quintero, M.C.; Gutierrez, E.E.; Sampayo, I.P.; Hernandez, J.R.; Rahman, M.H. Cable driven exoskeleton for upper-limb rehabilitation: A design review. *Robot. Auton. Syst.* **2020**, *126*, 103445. [CrossRef]
3. Pérez Vidal, A.F.; Rumbo Morales, J.Y.; Ortiz Torres, G.; Sorcia Vázquez, F.d.J.; Cruz Rojas, A.; Brizuela Mendoza, J.A.; Rodríguez Cerda, J.C. Soft Exoskeletons: Development, Requirements, and Challenges of the Last Decade. *Actuators* **2021**, *10*, 166. [CrossRef]
4. Chen, C.; Wu, X.; Liu, D.; Feng, W.; Wang, C. Design and Voluntary Motion Intention Estimation of a Novel Wearable Full-Body Flexible Exoskeleton Robot. *Mob. Inf. Syst.* **2017**, *2017*, 8682168. [CrossRef]

5. Serrancoli, G.; Falisse, A.; Dembia, C.; Vantilt, J.; Tanghe, K.; Lefeber, D.; Jonkers, I.; De Schutter, J.; De Groote, F. Subject-Exoskeleton Contact Model Calibration Leads to Accurate Interaction Force Predictions. *IEEE Trans. Neural Syst. Rehabil. Eng.* **2019**, *27*, 1597–1605. [CrossRef] [PubMed]
6. Zhang, X.; Li, H.; Lu, Z.; Yin, G. Homology Characteristics of EEG and EMG for Lower Limb Voluntary Movement Intention. *Front. Neurorobot.* **2021**, *15*, 642607. [CrossRef]
7. Saravanan, A.; Jangir, N.K.; Chakki, S. A Method to Control Bionic Arm Using Galvanic Skin Response. In Proceedings of the 8th International Conference on Communication Systems and Networks (COMSNETS), Bangalore, India, 5–10 January 2016.
8. Benabid, A.L.; Costecalde, T.; Eliseyev, A.; Charvet, G.; Verney, A.; Karakas, S.; Foerster, M.; Lambert, A.; Morinière, B.; Abroug, N.; et al. An exoskeleton controlled by an epidural wireless brain–machine interface in a tetraplegic patient: A proof-of-concept demonstration. *Lancet Neurol.* **2019**, *18*, 1112–1122. [CrossRef]
9. Li, Z.; Huang, Z.; He, W.; Su, C.-Y. Adaptive Impedance Control for an Upper Limb Robotic Exoskeleton Using Biological Signals. *IEEE Trans. Ind. Electron.* **2017**, *64*, 1664–1674. [CrossRef]
10. Pan, C.-T.; Chang, C.-C.; Yang, Y.-S.; Yen, C.-K.; Kao, Y.-H.; Shiue, Y.-L. Development of MMG sensors using PVDF piezoelectric electrospinning for lower limb rehabilitation exoskeleton. *Sens. Actuators A Phys.* **2020**, *301*, 111708. [CrossRef]
11. Li, Z.; Li, J.; Zhao, S.; Yuan, Y.; Kang, Y.; Chen, C.L.P. Adaptive Neural Control of a Kinematically Redundant Exoskeleton Robot Using Brain-Machine Interfaces. *IEEE Trans. Neural Netw. Learn. Syst.* **2019**, *30*, 3558–3571. [CrossRef]
12. Gordleeva, S.Y.; Lukoyanov, M.V.; Mineev, S.A.; Khoruzhko, M.A.; Mironov, V.I.; Kaplan, A.Y.; Kazantsev, V.B. Exoskeleton Control System Based on Motor-Imaginary Brain–Computer Interface. *Sovrem. Tehnol. V Med.* **2017**, *9*, 31–36. [CrossRef]
13. Trigili, E.; Grazi, L.; Crea, S.; Accogli, A.; Carpaneto, J.; Micera, S.; Vitiello, N.; Panarese, A. Detection of movement onset using EMG signals for upper-limb exoskeletons in reaching tasks. *J. Neuroeng. Rehabil.* **2019**, *16*, 45. [CrossRef] [PubMed]
14. Xu, H.; Xiong, A. Advances and Disturbances in sEMG-Based Intentions and Movements Recognition: A Review. *IEEE Sens. J.* **2021**, *21*, 13019–13028. [CrossRef]
15. Esposito, D.; Andreozzi, E.; Fratini, A.; Gargiulo, G.D.; Savino, S.; Niola, V.; Bifulco, P. A Piezoresistive Sensor to Measure Muscle Contraction and Mechanomyography. *Sensors* **2018**, *18*, 2553. [CrossRef]
16. Woodward, R.B.; Shefelbine, S.J.; Vaidyanathan, R. Pervasive Monitoring of Motion and Muscle Activation: Inertial and Mechanomyography Fusion. *IEEE/ASME Trans. Mechatron.* **2017**, *22*, 2022–2033. [CrossRef]
17. Su, H.; Hu, Y.; Karimi, H.R.; Knoll, A.; Ferrigno, G.; De Momi, E. Improved recurrent neural network-based manipulator control with remote center of motion constraints: Experimental results. *Neural Netw.* **2020**, *131*, 291–299. [CrossRef] [PubMed]
18. Qi, W.; Ovur, S.E.; Li, Z.; Marzullo, A.; Song, R. Multi-Sensor Guided Hand Gesture Recognition for a Teleoperated Robot Using a Recurrent Neural Network. *IEEE Robot. Autom. Lett.* **2021**, *6*, 6039–6045. [CrossRef]
19. Wu, H.; Huang, Q.; Wang, D.; Gao, L. A CNN-SVM combined model for pattern recognition of knee motion using mechanomyography signals. *J. Electromyogr. Kinesiol.* **2018**, *42*, 136–142. [CrossRef]
20. Shiman, F.; Lopez-Larraz, E.; Sarasola-Sanz, A.; Irastorza-Landa, N.; Spuler, M.; Birbaumer, N.; Ramos-Murguialday, A. Classification of different reaching movements from the same limb using EEG. *J. Neural. Eng.* **2017**, *14*, 046018. [CrossRef]
21. Xie, H.; Li, G.; Zhao, X.; Li, F. Prediction of Limb Joint Angles Based on Multi-Source Signals by GS-GRNN for Exoskeleton Wearer. *Sensors* **2020**, *20*, 1104. [CrossRef]
22. Gautam, A.; Panwar, M.; Biswas, D.; Acharyya, A. MyoNet: A Transfer-Learning-Based LRCN for Lower Limb Movement Recognition and Knee Joint Angle Prediction for Remote Monitoring of Rehabilitation Progress From sEMG. *IEEE J. Transl. Eng. Health Med.* **2020**, *8*, 2100310. [CrossRef] [PubMed]
23. Lu, L.; Wu, Q.; Chen, X.; Shao, Z.; Chen, B.; Wu, H. Development of a sEMG-based torque estimation control strategy for a soft elbow exoskeleton. *Robot. Auton. Syst.* **2019**, *111*, 88–98. [CrossRef]
24. Gui, K.; Liu, H.; Zhang, D. A Practical and Adaptive Method to Achieve EMG-Based Torque Estimation for a Robotic Exoskeleton. *IEEE/ASME Trans. Mechatron.* **2019**, *24*, 483–494. [CrossRef]
25. Ao, D.; Song, R.; Gao, J. Movement Performance of Human-Robot Cooperation Control Based on EMG-Driven Hill-Type and Proportional Models for an Ankle Power-Assist Exoskeleton Robot. *IEEE Trans. Neural Syst. Rehabil. Eng.* **2017**, *25*, 1125–1134. [CrossRef]
26. Wu, Q.; Chen, B.; Wu, H. Neural-network-enhanced torque estimation control of a soft wearable exoskeleton for elbow assistance. *Mechatronics* **2019**, *63*, 102279. [CrossRef]
27. Qi, W.; Aliverti, A. A Multimodal Wearable System for Continuous and Real-time Breathing Pattern Monitoring During Daily Activity. *IEEE J. Biomed. Health Inform.* **2020**, *24*, 2199–2207. [CrossRef]
28. Qi, W.; Wang, N.; Su, H.; Aliverti, A. DCNN based human activity recognition framework with depth vision guiding. *Neurocomputing* **2021**, *486*, 261–271. [CrossRef]
29. Dinh, B.K.; Xiloyannis, M.; Cappello, L.; Antuvan, C.W.; Yen, S.-C.; Masia, L. Adaptive backlash compensation in upper limb soft wearable exoskeletons. *Robot. Auton. Syst.* **2017**, *92*, 173–186. [CrossRef]
30. Dinh, B.K.; Xiloyannis, M.; Antuvan, C.W.; Cappello, L.; Masia, L. Hierarchical Cascade Controller for Assistance Modulation in a Soft Wearable Arm Exoskeleton. *IEEE Robot. Autom. Lett.* **2017**, *2*, 1786–1793. [CrossRef]
31. Little, K.; Antuvan, C.W.; Xiloyannis, M.; de Noronha, B.A.P.S.; Kim, Y.G.; Masia, L.; Accoto, D. IMU-based assistance modulation in upper limb soft wearable exosuits. In Proceedings of the 16th IEEE International Conference on Rehabilitation Robotics(ICORR), Toronto, ON, Canada, 24–28 June 2019; pp. 1197–1202.

32. Xiloyannis, M.; Chiaradia, D.; Frisoli, A.; Masia, L. Physiological and kinematic effects of a soft exosuit on arm movements. *J. Neuroeng. Rehabil.* **2019**, *16*, 29. [CrossRef]
33. Lessard, S.; Pansodtee, P.; Robbins, A.; Baltaxe-Admony, L.B.; Trombadore, J.M.; Teodorescu, M.; Agogino, A.; Kurniawan, S. CRUX: A compliant robotic upper-extremity eXosuit for lightweight, portable, multi-joint muscular augmentation. In Proceedings of the International Conference on Rehabilitation Robotics (ICORR), London, UK, 17–20 July 2017; pp. 1633–1638.
34. Lessard, S.; Pansodtee, P.; Robbins, A.; Trombadore, J.M.; Kurniawan, S.; Teodorescu, M. A Soft Exosuit for Flexible Upper-Extremity Rehabilitation. *IEEE Trans. Neural Syst. Rehabil. Eng.* **2018**, *26*, 1604–1617. [CrossRef] [PubMed]
35. Hosseini, M.; Meattini, R.; San-Millan, A.; Palli, G.; Melchiorri, C.; Paik, J. A sEMG-Driven Soft ExoSuit Based on Twisted String Actuators for Elbow Assistive Applications. *IEEE Robot. Autom. Lett.* **2020**, *5*, 4094–4101. [CrossRef]

electronics

Article

A Dual-Armed Robotic Puncture System: Design, Implementation and Preliminary Tests

Yongzhuo Gao, Xiaomin Liu, Xu Zhang, Zhanfeng Zhou, Wenhe Jiang, Lei Chen, Zheng Liu, Dongmei Wu and Wei Dong *

State Key Laboratory of Robotics and System, Harbin Institute of Technology, Harbin 150001, China; gaoyongzhuo@hit.edu.cn (Y.G.); liuxiaomin@stu.hit.edu.cn (X.L.); zhangxu950405@gmail.com (X.Z.); zhanfeng.zhou@mail.utoronto.ca (Z.Z.); jwh.eric@hit.edu.cn (W.J.); chenlei1991@stu.hit.edu.cn (L.C.); yanyao@stu.hit.edu.cn (Z.L.); wdm@hit.edu.cn (D.W.)

* Correspondence: dongwei@hit.edu.cn

Abstract: Traditional renal puncture surgery requires manual operation, which has a poor puncture effect, low surgical success rate, and high incidence of postoperative complications. Robot-assisted puncture surgery can effectively improve the accuracy of punctures, improve the success rate of surgery, and reduce the occurrence of postoperative complications. This paper provides a dual-armed robotic puncture scheme to assist surgeons. The system is divided into an ultrasound scanning arm and a puncture arm. Both robotic arms with a compliant positioning function and master–slave control function are designed, respectively, and the control system is achieved. The puncture arm's position and posture are decoupled by the wrist RCM mechanism and the arm decoupling mechanism. According to the independent joint control principle, the compliant positioning function is realized based on the single-joint human–computer interactive admittance control. The simulation and tests verify its functions and performance. The differential motion incremental master–slave mapping strategy is used to realize the master–slave control function. The error feedback link is introduced to solve the cumulative error problem in the master–slave control. The dual-armed robotic puncture system prototype is established and animal tests verify the effectiveness.

Keywords: puncture robot; renal puncture; master–slave control; admittance control

Citation: Gao, Y.; Liu, X.; Zhang, X.; Zhou, Z.; Jiang, W.; Chen, L.; Liu, Z.; Wu, D.; Dong, W. A Dual-Armed Robotic Puncture System: Design, Implementation and Preliminary Tests. *Electronics* **2022**, *11*, 740. https://doi.org/10.3390/electronics11050740

Academic Editors: João Paulo Morais Ferreira and Tao Liu

Received: 7 December 2021
Accepted: 22 February 2022
Published: 28 February 2022

Publisher's Note: MDPI stays neutral with regard to jurisdictional claims in published maps and institutional affiliations.

1. Introduction

A kidney biopsy is currently the main method that can achieve a clear diagnosis of diffuse renal diseases [1]. At present, there are two methods to achieve needle biopsy, namely the freehand puncture and ultrasound guided puncture. Freehand biopsy in the traditional way is labor-intensive. With the progression of the operation, the operation accuracy declines continuously. In addition, the inevitable tremor during the puncture operation greatly affects the puncture effect, leading to a low success rate of punctures. The operation is accompanied by mild or severe complications, which seriously endangers the health of the patient [2]. Thus, it is necessary to develop a robot-assisted percutaneous puncture system, to enable doctors to perform renal puncture operations more accurately and efficiently and further reduce the pain of the patient during the operation.

Due to the strong clinical demand for puncture surgery, many institutions have carried out related research. Stoianovici et al. [3] developed a spinal and renal puncture surgery robot named "Acubot", and a pneumatically driven puncture surgery robot that can replace the puncture mechanism [4]. Mitchell et al. [5] developed a retinal surgery robot. Üneri et al. [6] and He et al. [7] improved its human–computer interaction performance and obtained EyeRoBot2 and EyeRoBot2.1, respectively. The navigation of puncture surgery robots mainly uses external image guidance [8], including CT [9,10], MRI [11,12], and ultrasound [13–16], among which ultrasound image guidance is widely used. Boctor et al. [13] developed an ultrasound-guided hepatic puncture surgical robot. Hong et al. [14] developed an ultrasound-guided puncture robot, UMI, which can adjust the puncture needle path through real-time ultrasound navigation. Kettenbach et al. [15] designed a tissue biopsy puncture robot, B-Robot-I, based on ultrasound navigation. Bassan et al. [16] developed a three-dimensional ultrasound navigation robot for prostate tumor puncture surgery. Although the puncture robot has been studied for years, it lacks practicality. A very important reason is that the existing system is unable to meet the safety requirement of surgery and satisfy the usage habits of surgeons.

The clinical demand for robot-assisted surgery has also promoted the research of surgical robot systems. Computer Motion has developed the ZEUS surgical robot system [17,18]. Intuitive Surgical has developed the master–slave teleoperation da Vinci surgical robot system [19]; after system improvement, the da Vinci Xi [20] and da Vinci SP systems have been developed, whose effectiveness is widely accepted. Hannaford et al. [21] developed the open-source surgical system Raven II. Konietschke et al. [22] and Hagn et al. [23] developed the DLR MiroSurge lightweight surgical robot system. Although most robot-assisted surgery robots interpret the meaning of minimally invasive, commercial systems pay more attention to the preoperative docking effect and the master–slave operation during the operation. In a sense, whether docking and master–slave operation are easy to use determines whether the puncture robot can be applied to clinical applications.

At present, there are few clinical applications on the robotic system for renal puncture. The main contributions of this paper are as follows. (1) A dual-armed robotic system prototype with master–slave control and a dragging docking function for percutaneous puncture is reported to assist the traditional renal puncture. The system is divided into an ultrasound scanning arm and a puncture arm. The puncture arm is designed with a position and posture decoupling function to guarantee the puncture safety. (2) An interaction force control method is proposed for the positioning arm during the docking operation. (3) The robotic system for renal puncture is verified by animal tests.

This paper is organized as follows. Section 2 describes the mechanical design and the kinematic modeling of the dual-armed robotic system. Section 3 presents the robot control system and control strategy. Section 4 describes the simulation analysis of the control strategy and the animal tests. In the last part, the designed dual-armed robotic system for renal puncture is summarized and prospected.

2. System Design and Modeling

The system architecture of the dual-armed robotic system is shown in Figure 1. The dual-armed robotic system is divided into an ultrasonic (US) scanning arm and a puncture arm. The ultrasonic robotic arm is designed to provide image guidance for the operation. The puncture arm with the position and posture decoupling function performs the puncture operation, with the consideration of the puncture safety. During the docking operation, the two arms are positioned on the patient by dragging. During the surgery, the surgeons can use one master operator to control the US arm moving on patients for scanning, while using another master operator to control the puncture arm to adjust the posture of the needle.

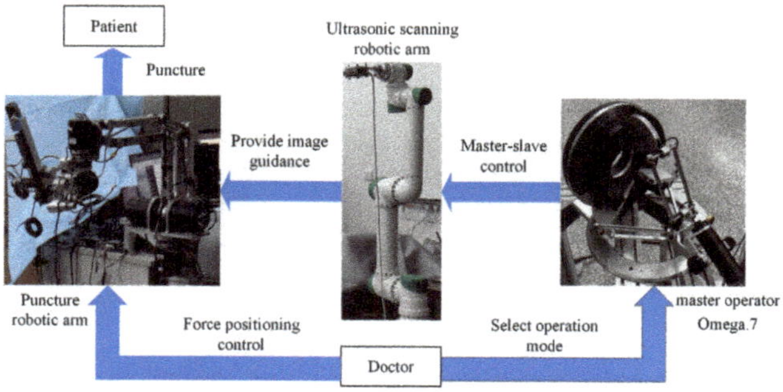

Figure 1. Dual-arm puncture robot system architecture.

2.1. *Mechanical Design and Kinematic Modeling of Puncture Arm*

2.1.1. Mechanical Design of Puncture Arm

During the dragging process, the robotic arm is easily restricted by the flexible working space and joint singularities of the robotic arm. In order to make the dragging process simple and easy, the puncture arm adopts a position and posture decoupling mechanical structure.

The puncture arm is divided into the arm and wrist. The arm and wrist determine the position and posture of the puncture needle, respectively. Regardless of the rotation of the puncture needle, the puncture arm only needs five degrees of freedom to complete the positioning of the puncture needle. However, redundant degrees of freedom are required for doctors to avoid some surgical positions. Thus, we provide a redundant degree of freedom at the end of the arm. The wrist provides two posture angles for the puncture needle. The arm is decoupled from the wrist, and the arm is responsible for giving the position of the puncture point and the redundant posture angle. The arm structure is shown in Figure 2a. The wrist adopts a two-dimensional Remote Center of Motion (RCM) mechanism (see Aksungur [24]), and its distal virtual point is the required puncture point. The range of motion of the two-dimensional RCM mechanism can be calculated by the required puncture angle. The arm is actually a decoupling mechanism. The RCM mechanism used in this paper is shown in Figure 2b. The final mechanical structure of the puncture arm is shown in Figure 2c.

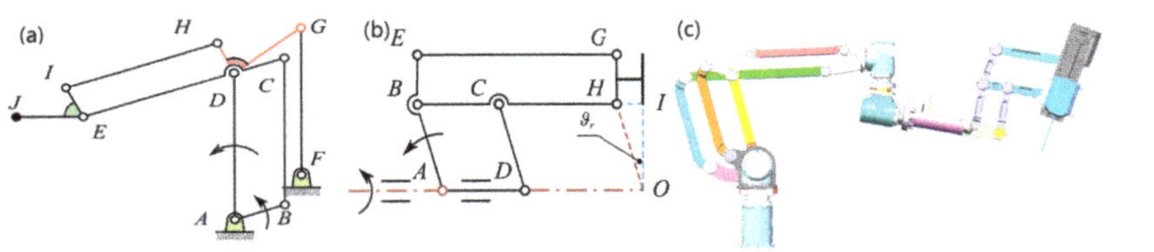

Figure 2. Puncture arm structure design. (**a**) Arm decoupling mechanism; (**b**) Wrist RCM mechanism; (**c**) Puncture arm structure.

2.1.2. Kinematic Modeling of the Puncture Arm

Due to the existence of the arm decoupling mechanism, a passive joint is introduced. Thus, the puncture arm actually contains seven joints with kinematic meaning but has only six degrees of freedom. The D-H coordinate system is established for the mentioned seven joints, as shown in Figure 3. Among them, joint $J_{3'}$ is the redundant passive degree of freedom caused by position and posture decoupling and joint 7 represents the RCM point at the end. The positive kinematics equation contains seven joint variables and defines six generalized variables $q_1 - q_6$ corresponding to motors. The D-H parameters are expressed by generalized variables, as shown in Table 1.

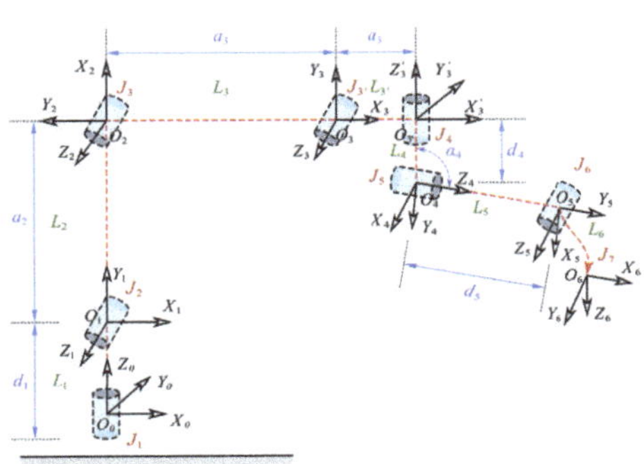

Figure 3. D-H coordinate system of the puncture arm.

Table 1. D-H parameters of the puncture arm expressed by generalized variables.

Link	1	2	3	3′	4	5	6
θ_i	q_1	$q_2 + 90°$	$q_3 - q_2 - 90°$	$-q_3$	$q_4 - 90°$	$q_5 + 90°$	$q_6 + 90°$

2.2. Mechanical Design and Kinematic Modeling of Scanning Arm

2.2.1. Mechanical Design of Scanning Arm

The function of the scanning arm is to provide image guidance for the puncture operation. High-quality images are conducive to the improvement of the accuracy of the operation. Thus, it is necessary to have different operation modes for doctors to choose. However, it has no special requirements for the structure design, so the choice of the scanning arm is a common 6-degree-of-freedom universal robotic arm with an ultrasonic probe, as shown in Figure 4a. These six joints are all rotary joints, and the axes of joints 2, 3, and 4 are parallel to each other, which has good working space and operation dexterity.

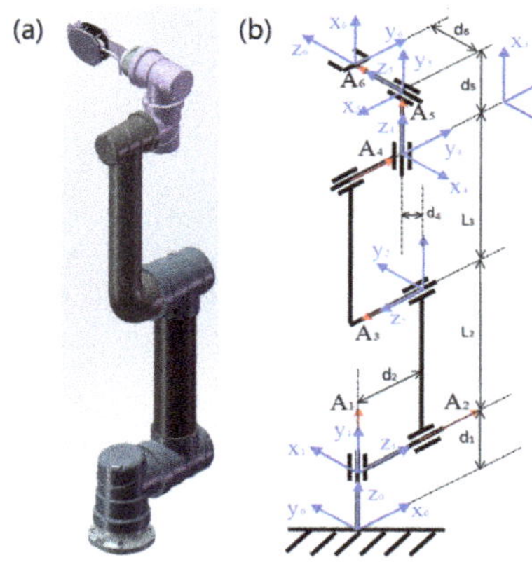

Figure 4. The structure and D-H coordinate system of the scanning arm. (**a**) Scanning arm structure; (**b**) Scanning arm link coordinate system.

2.2.2. Kinematic Modeling of Scanning Arm

The six joints of the scanning arm are all designed as rotating joints. The linkage parameters and D-H coordinate system of the scanning arm are shown in Figure 4b. The positive kinematics equation contains six joint variables, and the D-H parameters expressed by joint variables are shown in Table 2.

Table 2. D-H parameters of the scanning arm represented by joint variables.

Link	1	2	3	4	5	6
θ_i	θ_1 (+90°)	θ_2 (+90°)	θ_3	θ_4 (+90°)	θ_5 (−90°)	θ_6 (+90°)

3. The Implementation of the Control System

The dual-armed puncture system adopts a centralized–distributed architecture, as shown in Figure 5. The upper-level control algorithm is centralized to the industrial computer to realize, and the joint motor servo control algorithm is distributed to the independent driver on the bus to realize. The upper layer and the bottom layer are linked together through the connection layer, including the software interface that connects the PLC and Numerical Control (NC), and the hardware interface including an EtherCAT coupler and general IO module.

<table>
<tr><td>Control Layer</td><td>Centralized Control</td><td>Motion Control Algorithm</td><td>IPC</td></tr>
<tr><td>Connection Layer</td><td colspan="2">Software Connection (PLC-NC)</td><td>Hardware Connection (EtherCAT Interface, IO Module)</td></tr>
<tr><td>Physical Layer</td><td>Distributed Control</td><td>Joint Motor Servo Control Algorithm</td><td>Motor Driver</td></tr>
</table>

Figure 5. Dual-arm puncture system control architecture.

3.1. Master–Slave Control Strategy

This paper adopts the master–slave mapping strategy of differential motion increment based on the inverse Jacobian matrix, and the master hand uses Omega7 of the Force Dimension company. The master–slave mapping strategy uses the differential displacement increment in a small time period to replace the instantaneous speed, and the linear solution can be obtained only by using the Jacobian matrix, which is convenient for programming and calculation.

The linear solution equation for the discretized scanning arm inverse kinematics problem can be described as:

$$\begin{cases} \Delta q = J(q)^{-1}\Delta x_e \\ \Delta q = \dot{q}\Delta t \\ \Delta x_e = \begin{bmatrix} \dot{p}_e & \dot{\phi}_e \end{bmatrix}^T \Delta t \end{cases} \tag{1}$$

where Δq is the differential increment of the joint space, Δx_e is the differential motion increment of the position and Euler angle of the end of the scanning arm in Cartesian space, and the time interval is Δt.

In addition, this method of linearly equivalent generalized velocity to differential motion increment has the following errors:

$$e = \Delta x_e - J(q)\Delta q. \tag{2}$$

The error e is a high-order error and can be ignored in the calculation. However, this kind of small error will continue to accumulate, leading to a large master–slave following error of the end effector. Thus, the differential motion incremental master–slave mapping algorithm based on the inverse Jacobian matrix needs to introduce a feedback link, to eliminate the accumulation of errors.

The master–slave real-time control algorithm in this paper is as follows. Collecting the differential motion increment of the position and posture of the master hand Omega7, the working space of the master hand and scanning arm are combined to determine the master–slave space mapping function. Then, the expected differential motion increment of the end effector's position and posture of the scanning arm is obtained through mapping. After this, the inverse Jacobian matrix is used to calculate the expected differential angle increment of each joint of the scanning arm, and finally the expected angle of each joint is calculated. Lastly, the expected angle of each joint is used as the input of the underlying control algorithm of the joint motor module.

The master–slave control strategy based on the differential motion increment of the inverse Jacobian matrix with error compensation is shown in Figure 6. The control process uses speed as a variable, but in the actual algorithm, the differential motion increment is used for programming calculations instead of speed. This kind of master–slave control algorithm reduces the calculation time, and the design of the error feedback link also improves the accuracy of the master–slave control algorithm.

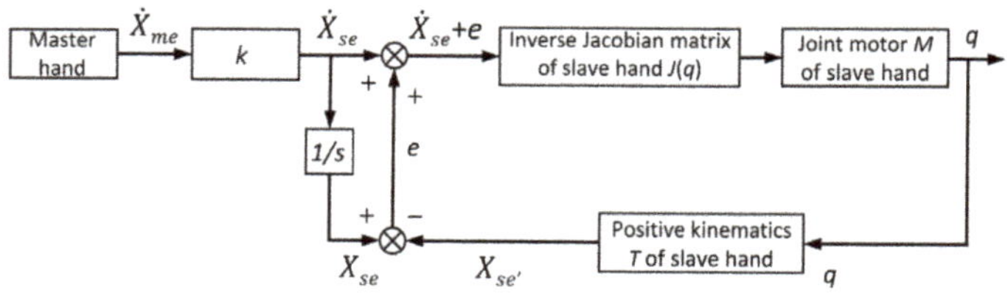

Figure 6. Master–slave control of inverse Jacobian matrix with error compensation.

3.2. Control Strategy for Positioning during Docking

The positioning control strategy can realize the preoperative positioning of the arm during docking operation. The surgeons can drag the arm directly to determine the position and posture of the puncture operation. In this situation, the arm needs to provide a compliant operating experience. According to the characteristics of the puncture arm and the architecture of the control system, this paper proposes a joint-space admittance control strategy to achieve compliant positioning operation, which is based on the admittance control strategy for independent joint control. This method actually superimposes the admittance control to an independent joint control strategy, to obtain the compliant control effect. During the drag process of the puncture arm, it will be affected by the gravity moment and the nonlinear coupling moment. In the drag application, the acceleration is small enough, and the nonlinear coupling moment can generally be ignored. However, the gravity moment cannot be ignored. In this paper, joint torque sensors are installed for each drive joint for gravity moment identification.

3.2.1. Design of Mass Damping Controller

The principle of the human–computer interaction admittance control is shown in Figure 7. The interaction force acts on each joint, so there is no need for the transformation of the transposed Jacobian matrix. In the actual design of the controller, it can be considered that the output of the admittance controller is a speed command. The admittance of the controller [25] is defined as

$$Y = \frac{\dot{q_d}}{F_{ext}}. \tag{3}$$

In the controller, admittance Y only represents the conversion from drag force F_{ext} to control speed $\dot{q_d}$.

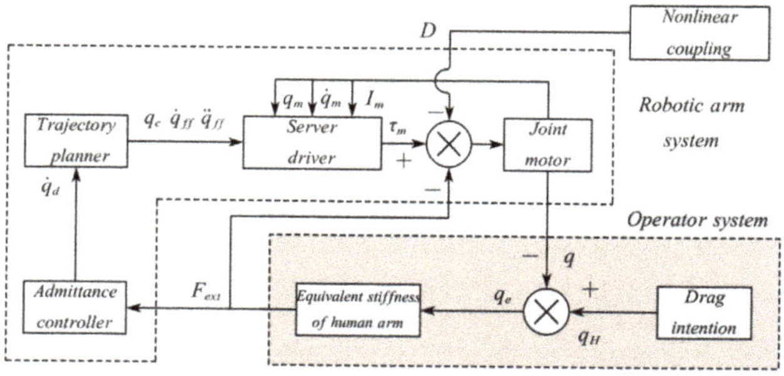

Figure 7. Human–computer interaction admittance control principle.

The most ideal equal admittance control system of the robotic arm joint is a mass damping system. The process of dragging a robotic arm joint equals dragging a mass. The mass and damping of the mass and the equal stiffness of the human arm determine the dragging experience and effect. The equivalent control block diagram of the human–computer interaction system considering the operator is shown in Figure 8.

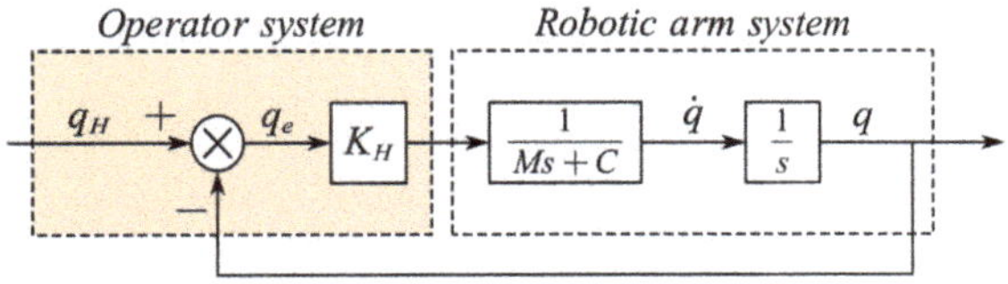

Figure 8. Equivalent control block diagram of human–computer interaction system.

A simple approximation is made here. The operating system (the human) is equivalent to a fixed stiffness K_H. Thus, the equivalent model of the entire system is linearized as a second-order system, which can be described as

$$Y = \frac{K_H}{Ms^2 + Cs + K_H}, \tag{4}$$

when the admittance controller selects the mass damping system, expressed as

$$Y = \frac{\dot{q}_d}{F_{ext}} = \frac{1}{Ms + C}. \tag{5}$$

The system represented by Equation (4) can be obtained. The second-order system is the simplest form of the control system, and the desired effect can be easily achieved by adjusting the parameters. However, this situation is not in line with reality, so the purpose of the admittance controller is to choose a suitable controller architecture, to make the system as close to the ideal form as possible.

3.2.2. Design of Human–Computer Interactive Admittance Controller

The human–computer interaction system shown in Figure 7 is actually nonlinear. The nonlinearity of the operator's arm system [26] is difficult to describe with an accurate mathematical model, but a reasonable approximation can be made, assuming that the operator's arm is a variable stiffness impedance system, which can be linearized and approximated. The result after the linear approximation is shown in Figure 9. In the linearized model, the factors affecting the human–computer interaction performance mainly include three aspects: the stiffness of the human arm K_A, the designed admittance controller Y, and the servo drive models G and S. The control of the linearized model is actually a series correction of the system. In fact, when the admittance controller adopts the lag correction strategy, the best effect is achieved. The typical lag correction strategy is as follows:

$$Y = \frac{\dot{q}_d}{F_{ext}} = K_Y \frac{T_1 s + 1}{T_2 s + 1}. \tag{6}$$

After adding the operator system and admittance controller to form a new system, system stability should be reconsidered. The stability of the system is mainly affected by the stiffness K_A of the human arm and the admittance controller Y. The arm stiffness K_A and the gain K_Y in the admittance controller can be equal to the coefficient K for consideration.

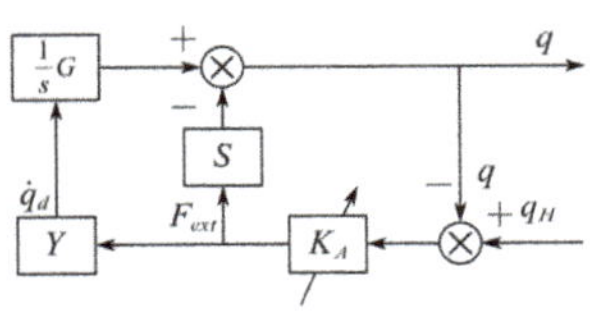

Figure 9. Human–computer interactive admittance control system after linearization approximation.

3.3. Joint Gravity Moment Modeling and Parameter Identification

Since the puncture arm has seven degrees of freedom, there are seven gravity moment parameters for seven generalized variables q. As shown in Figure 10, all the link combinations in kinematics are given, where $L1x$ is the length of links for axis 1, $L2x$ for axis 2 and so on. Obviously, gravity moments g_1 and g_4 for q_1 and q_4 are zero. As the position and posture are decoupled, the motions of joint 2 and joint 3 will not change the posture of joints 5, 6, and 7. Thus, we can make the following simplification:

$$\begin{aligned} g_1(q) &= g_4(q) = 0 \\ g_2(q) &= ac_2 + bs_2 \\ g_3(q) &= dc_3 + es_3, \end{aligned} \tag{7}$$

where c_2 denotes $cos(q_2)$, s_2 denotes $sin(q_2)$. a, b, d, e denote the parameters of the corresponding links. Similarly, for g_5, g_6 and g_7,

$$\begin{aligned} g_5(q) &= \boldsymbol{A}^T[c_5 \quad s_5 \quad s_5c_6 \quad s_5s_6 \quad d_7s_5c_6]^T \\ g_6(q) &= \boldsymbol{B}^T[s_6 \quad c_6 \quad c_5s_6 \quad c_5c_6 \quad d_7c_6 \quad d_7c_5s_6]^T \\ g_7(q) &= \boldsymbol{C}^T[s_6 \quad c_5c_6]^T, \end{aligned} \tag{8}$$

where d_7 denotes the displacement of needle feeding motion on axis 7. Moreover, $\boldsymbol{A} \in \mathbb{R}^5$, $\boldsymbol{B} \in \mathbb{R}^6$, and $\boldsymbol{C} \in \mathbb{R}^3$ are parameter vectors to be identified.

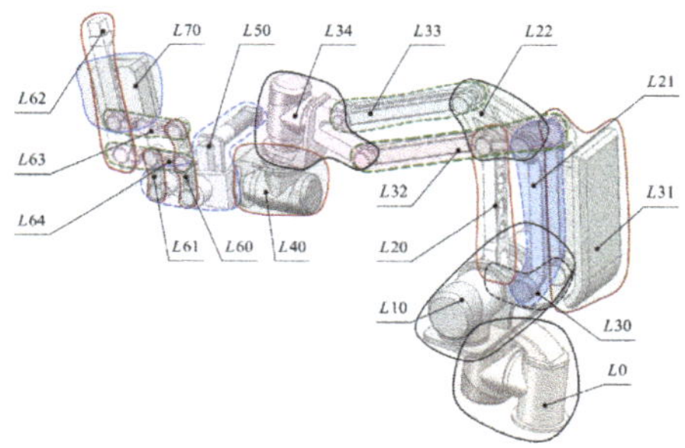

Figure 10. The link combinations in kinematics.

For parameter identification, the least squares method is adopted. Take joint 5 as an example, define input variable $\boldsymbol{x} = [c_5 \quad s_5 \quad s_5c_6 \quad s_5s_6 \quad d_7s_5c_6]^T$ and output variable $y = g_5(\boldsymbol{q})$, parameter vector $\boldsymbol{A}$. The vector formula is expressed as

$$y = \boldsymbol{x}^T\boldsymbol{A} + e, \tag{9}$$

where $\boldsymbol{e}$ is the residual errors. $N(N > 5)$ times of identification shall be carried out to obtain vectors $\boldsymbol{X}$, $\boldsymbol{Y}$, and $\boldsymbol{E}$. Apply the least square method to find the vector $\boldsymbol{A}$.

$$\boldsymbol{A} = (\boldsymbol{X}^T\boldsymbol{X})^{-1}\boldsymbol{X}^T\boldsymbol{Y}, \tag{10}$$

The above analysis has obtained the identification method of the gravity moment of all generalized joint variables. During the identification process, the robot arm needs to keep moving at a low speed, and the value read by the joint torque sensor can be regarded as the joint gravity moment. After obtaining the identification parameters, during the force setting and dragging process, the data of the torque sensor are subtracted from the calculated joint gravitational moment to obtain the human–computer interaction torque. Since the configuration of the US arm is conventional, the parameter identification method will not be repeated here.

4. System Simulations and Animal Tests

During a renal puncture operation, two stages are performed for the dual-armed puncture system. We call them "docking" and "surgery". During the docking process, the operator positions the two arms at the patient's kidney and needle entry point, respectively. During the surgery process, master–slave control is the main operation for the puncture system. In this section, the authors verify the master–slave control motion and arm positioning during docking by simulation. Then, a cycled reciprocating motion of dragging test is carried out for verification. In the end, an animal renal puncture test is reported.

4.1. Master–Slave Control Algorithm Simulation

To form a master–slave control system, the algorithms are respectively packaged in the form of functional blocks and integrated into the TwinCAT software control system. Select each joint angle of the current position of the scanning arm as $\theta_1 = 90°$, $\theta_2 = 75°$, $\theta_3 = 90°$, $\theta_4 = 105°$, $\theta_5 = -90°$, $\theta_6 = 90°$, and use the randomly generated incremental sequence as the differential motion increment of the master hand to simulate the motion of the master hand. Then, the movement of the master hand is mapped to the end of the scanning arm through the master–slave mapping function, and the inverse Jacobian matrix is used to calculate the joint control amount. The master–slave control algorithm with or without an error feedback link is used to calculate the following error. The master–slave following error comparison of the two algorithms is shown in Figure 11. This algorithm has an obvious effect on reducing the following error.

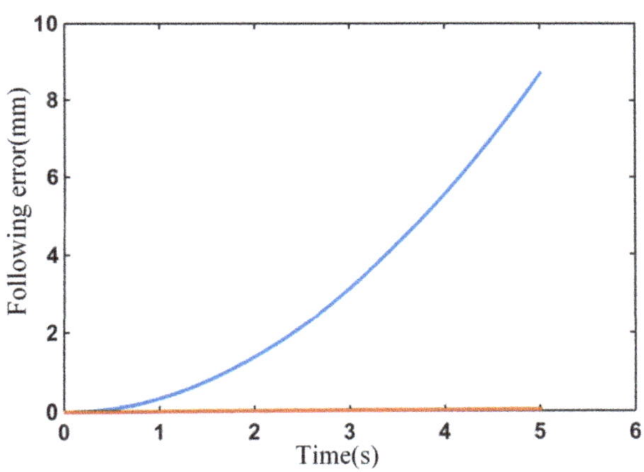

Figure 11. Comparison of following error with (red line) and without (blue line) error feedback link.

In Figure 11, the following error generated by the master–slave control algorithm without the error feedback link is significantly greater than the result after adding the error feedback link. Moreover, the following error generated by the master–slave control algorithm without error feedback accumulates with the movement of the master hand. Therefore, the faster the master hand, the longer the movement time, and the larger the following error obtained by the master–slave control algorithm without error feedback. In the simulation, the given speed of the main hand is large, which means that the following error generated is also large. The error gradually accumulates and reaches 8.7 mm finally, which is unacceptable for the surgical robot. After adding the error feedback link, the error accumulation situation is greatly improved, and the maximum accumulated error is reduced from 8.7 mm to 0.04 mm, which has a significant error reduction effect.

4.2. Simulation of Admittance Controller for Arm Positioning

The admittance controller is divided into the gain part and zero pole part. The gain in the admittance controller can be combined with the equivalent stiffness of the arm for analysis. Adjusting the gain of the admittance controller can be regarded as changing the stiffness of the arm. The output of the admittance controller is the speed command, which is used as the feedforward input of the servo controller. Its integral and derivative can be respectively used as the position and acceleration feedforward input. This is actually a continuous system simulation of the trajectory planner.

First, consider the case where the admittance controller is pure gain. During the simulation process, the stiffness of the arm is fixed to 100. When the gain of the admittance controller is changed, it can be considered that the stiffness of the arm changes. Figure 12 shows the case where the admittance controller is pure gain. It can be seen from the figure that when the gain increases, the system's low frequency gain and shear frequency will also increase, which means that the response speed and tracking accuracy of the system will increase. Moreover, the high tracking accuracy will reduce the interaction force of the system, thereby improving the dragging experience. However, the gain cannot be increased indefinitely; high gain will make the system unstable.

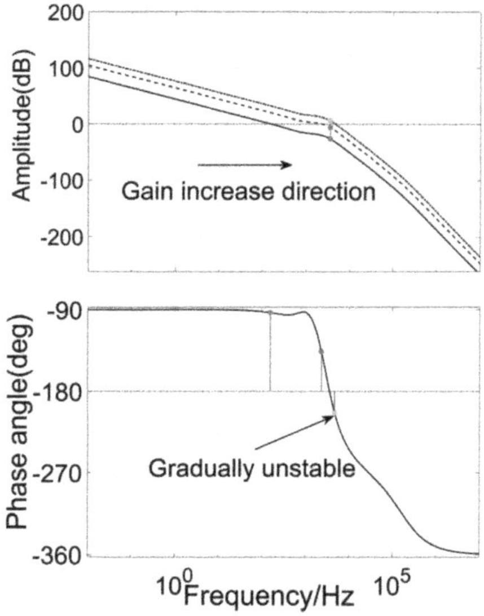

Figure 12. Admittance controller in the case of pure gain.

When the mass damping controller is introduced as the admittance controller, it needs to be discussed in two cases of stable gain and unstable gain. As shown in Figure 13a, when the gain part of the admittance controller stabilizes the system, and the mass damping controller is selected for the zero-pole part, it is equal to introducing a pole. If the pole is located after the shear frequency, it will not affect the low-frequency part of the system, and a small quality parameter needs to be set to obtain a larger pole. If the pole is located before the shear frequency, it will adversely affect the system, reducing the shear frequency and stability margin of the system, and it may even make the system unstable in severe cases. When the gain or arm stiffness is too large and the system is unstable, as shown in Figure 13b, the system can be stabilized by introducing a mass damping controller. This is actually obtaining the leading phase angle by sacrificing the shear frequency of the system, but one pole also causes the lag of the phase angle. The effects of the two cancel out each other, and it is difficult to obtain a higher stability margin.

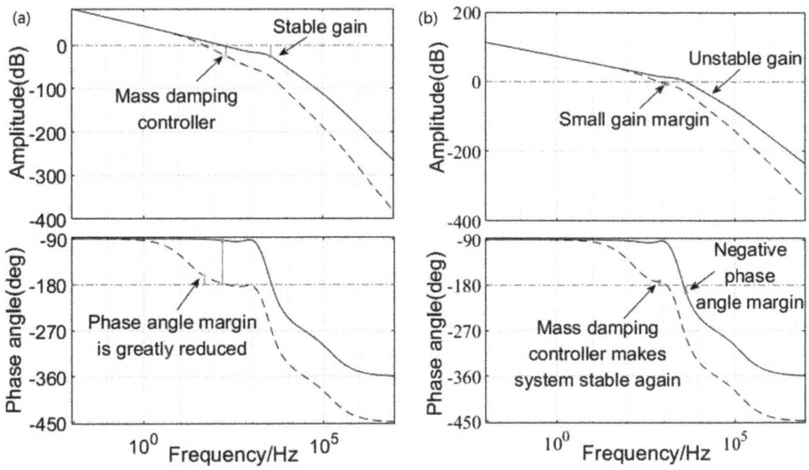

Figure 13. Analysis of the introduction of mass damping controller. (**a**) Stable gain introduces mass dampling controller; (**b**) Unstable gain introduces mass dampling controller.

In fact, the mass damping controller is equal to a low-pass filter for the interactive force signal. In the actual robotic arm, the interaction force is obtained by subtracting the gravity from the force sensor information, and it will be influenced by the low-frequency vibration interference of the robotic arm and the high-frequency interference of the electrical signal. The force sensor acquisition card can easily filter out high-frequency interference, and the low-frequency mechanical vibration needs to be filtered a second time. Thus, the bandwidth of the selected filter will be very low, which is equal to the pole of the mass damping controller being very small, and this will harm the system. In the above analysis, it is found that the effect of the single-pole mass damping controller is not obvious. In fact, the control of the linearized human–computer interactive admittance control system can be regarded as a problem of series system correction.

When the high gain makes the system unstable, the introduction of a lag corrector can work well, as shown in Figure 14. Lag correction is different from the single-pole mass damping controller, which can bring a larger phase angle margin through the additional zero pole points. After applying the lag correction, the system can tolerate a large system gain, which means that it can accept a greater arm stiffness.

Finally, it is found that the decisive factors for the force positioning function are actually the joint kinetic features of the robotic arm and the performance of the servo driver. The admittance controller is only used as a system correction device. For example, the following performance of the force positioning control depends on the maximum acceleration and deceleration capacity of the robotic arm joints. The admittance controller can only be used to ensure the stability of the system when the stiffness of the human arm changes drastically.

4.3. Cycled Reciprocating Motion of Dragging Tests

After comparison of the admittance controller with pure gain and with a lag corrector, cycled reciprocating motion of dragging tests are carried out. This work uses an open-loop admittance controller, with parameters $d = 5$ Ns/m, $m = 1$ kg. The two tests are situations of high dynamic performance and high sensitivity. The methods and results are shown in Figures 15 and 16.

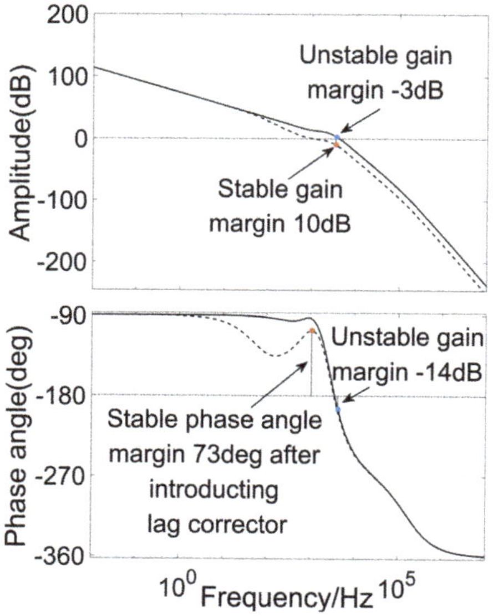

Figure 14. Analysis of lag compensation as admittance controller.

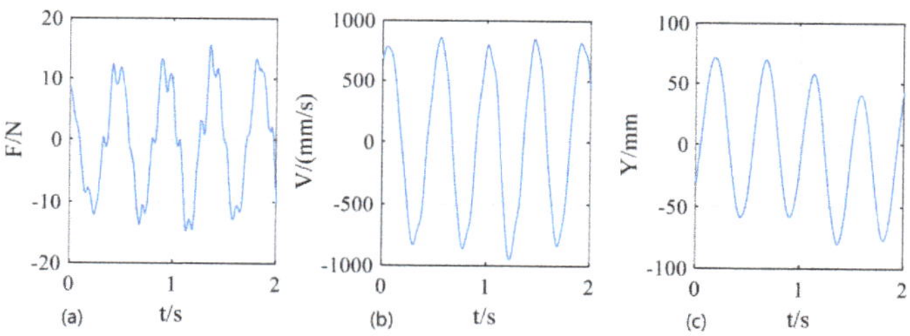

Figure 15. High dynamic performance motion test of dragging operation: (**a**) contact force; (**b**) robot velocity; (**c**) robot position.

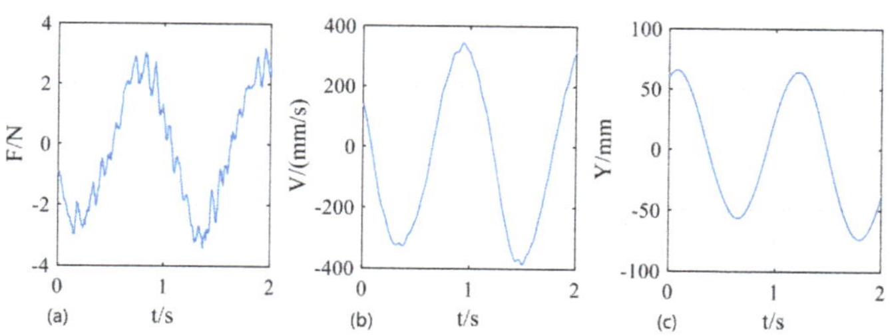

Figure 16. High sensitivity motion test of dragging operation: (**a**) contact force; (**b**) robot velocity; (**c**) robot position.

Figure 15 shows robot performance data in high dynamic performance and high stability. It can be seen that when the positioning amplitude is 50 mm, the positioning frequency reaches 2 Hz, the contact force does not exceed 15 N, and the maximum positioning speed in Cartesian space is as high as 800 mm/s. The data are stable and there is no divergence. In Figure 16, under low-speed conditions, the positioning amplitude is 50 mm, and the robot speed does not exceed 400 mm/s. The contact force does not exceed 3 N. The robot exhibits high sensitivity.

4.4. Animal Tests

Several animal renal puncture tests are carried out using the dual-armed robotic puncture system. Figure 17 shows the robotic puncture system prototype, including a US arm, a puncture arm, an operating console, and an ultrasonic monitor. Figure 18 shows the arm positioning operation (a and c) and master–slave control (b). Figure 18d shows the surgery process of the animal test. The animal tests verify the effectiveness of the dual-armed robotic puncture system.

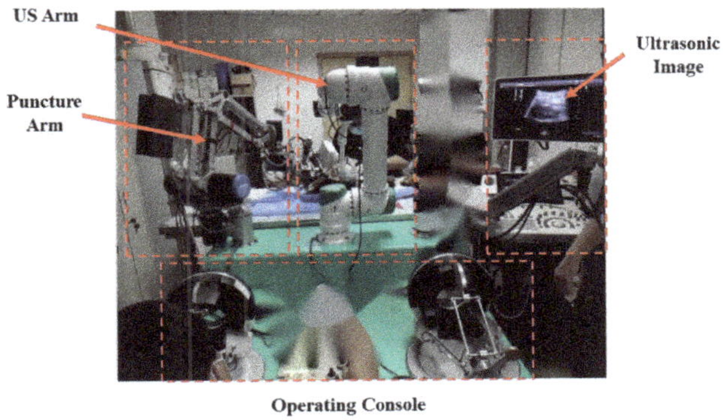

Figure 17. The prototype of dual-armed robotic puncture system and system composition.

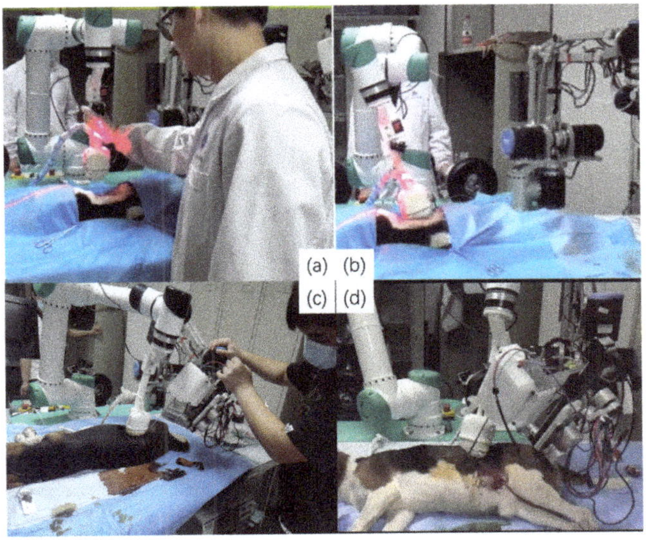

Figure 18. Process of the animal tests. (**a**) US arm docking operation; (**b**) master–slave control of US arm for image guidance; (**c**) puncture arm docking operation; (**d**) the surgery process of the animal test.

5. Conclusions

This paper proposes a dual-armed robotic puncture system for renal puncture. For the docking process, surgeons can position the arms by dragging. For the surgery process, surgeons can use the master operator to control the arms for image guidance and adjust the needle posture, respectively. This paper describes the mechanical design, control system design, and animal tests of the dual-armed robotic puncture system, which consists of a US arm that provides ultrasound guidance and a puncture arm with a position and posture decoupling function. The control strategy of each robotic arm has been selected according to the needs of puncture surgery and verified by a simulation and tests. The results show that the dual-armed robotic puncture system in this paper is easy to operate during the docking process, and the master–slave control can meet the demand of image guidance and puncture operations. The animal tests are reported to verify the effectiveness of the system prototype.

In the future, the dual-armed robotic puncture system can be used for clinical trials, which is expected to improve the kidney retreat and the up-and-down displacement caused by breathing during the puncture process, reducing the degree of rupture of the renal parenchyma and renal capsule. This will increase the success rate of puncture surgery and reduce the occurrence of complications, so the system has good application prospects.

Author Contributions: Conceptualization, D.W.; methodology, Y.G., X.L., X.Z. and Z.Z.; software, X.L.; validation, L.C. and Z.L.; writing—original draft preparation, W.J. and Y.G.; writing—review and editing, Y.G.; project administration, W.D. All authors have read and agreed to the published version of the manuscript.

Funding: This research was funded by the National Key Research and Development Program of China (Grant No. 2017YFB1303101).

Informed Consent Statement: The study was conducted according to the guidelines issued by the Ethics Committee of Huazhong University of Science and Technology (Project Partner) for National Key Research and Development Program of China 2017YFB1303101.

Data Availability Statement: Not applicable.

Conflicts of Interest: The authors declare no conflict of interest.

References

1. Jun, J.; Yaqing, C.; Yongchang, Z. Analysis of complications in ultrasound-guided percutaneous renal biopsy. *Chin. J. Ultrasound Med.* **2006**, *22*, 858–860.
2. Whittier, W.L.; Korbet, S.M. Timing of complications in percutaneous renal biopsy. *J. Am. Soc. Nephrol.* **2004**, *15*, 142–147. [CrossRef] [PubMed]
3. Stoianovici, D.; Cleary, K.; Patriciu, A.; Mazilu, D.; Stanimir, A.; Craciunoiu, N.; Watson, V.; Kavoussi, L. AcuBot: A robot for radiological interventions. *IEEE Trans. Robot. Autom.* **2003**, *19*, 927–930. [CrossRef]
4. Stoianovici, D.; Jun, C.; Lim, S.; Li, P.; Petrisor, D.; Fricke, S.; Sharma, K.; Cleary, K. Multi-imager compatible, MR safe, remote center of motion needle-guide robot. *IEEE Trans. Biomed. Eng.* **2017**, *65*, 165–177. [CrossRef] [PubMed]
5. Mitchell, B.; Koo, J.; Iordachita, I.; Kazanzides, P.; Kapoor, A.; Handa, J.; Hager, G.; Taylor, R. Development and application of a new steady-hand manipulator for retinal surgery. In Proceedings of the 2007 IEEE International Conference on Robotics and Automation, Rome, Italy, 10–14 April 2007; pp. 623–629.
6. Üneri, A.; Balicki, M.A.; Handa, J.; Gehlbach, P.; Taylor, R.H.; Iordachita, I. New steady-hand eye robot with micro-force sensing for vitreoretinal surgery. In Proceedings of the 2010 3rd IEEE RAS & EMBS International Conference on Biomedical Robotics and Biomechatronics, Tokyo, Japan, 26–29 September 2010; pp. 814–819.
7. He, X.; Roppenecker, D.; Gierlach, D.; Balicki, M.; Olds, K.; Gehlbach, P.; Handa, J.; Taylor, R.; Iordachita, I. Toward clinically applicable steady-hand eye robot for vitreoretinal surgery. In Proceedings of the ASME International Mechanical Engineering Congress and Exposition, Houston, TX, USA, 9–15 November 2012; Volume 45189, pp. 145–153.
8. Kettenbach, J.; Kronreif, G. Robotic systems for percutaneous needle-guided interventions. *Minim. Invasive Ther. Allied Technol.* **2015**, *24*, 45–53. [CrossRef] [PubMed]
9. Zhou, Y.; Thiruvalluvan, K.; Krzeminski, L.; Moore, W.H.; Xu, Z.; Liang, Z. CT-guided robotic needle biopsy of lung nodules with respiratory motion—Experimental system and preliminary test. *Int. J. Med. Robot. Comput. Assist. Surg.* **2013**, *9*, 317–330. [CrossRef] [PubMed]
10. Hungr, N.; Bricault, I.; Cinquin, P.; Fouard, C. Design and validation of a CT-and MRI-guided robot for percutaneous needle procedures. *IEEE Trans. Robot.* **2016**, *32*, 973–987. [CrossRef]
11. Patel, N.A.; Azimi, E.; Monfaredi, R.; Sharma, K.; Cleary, K.; Iordachita, I. Robotic system for MRI-guided shoulder arthrography: Accuracy evaluation. In Proceedings of the 2018 International Symposium on Medical Robotics (ISMR), Atlanta, GA, USA, 1–3 March 2018; pp. 1–6.
12. Stoianovici, D.; Kim, C.; Srimathveeravalli, G.; Sebrecht, P.; Petrisor, D.; Coleman, J.; Solomon, S.B.; Hricak, H. MRI-Safe Robot for Endorectal Prostate Biopsy. *IEEE/ASME Trans. Mechatron.* **2014**, *19*, 1289–1299. [CrossRef]
13. Boctor, E.M.; Fischer, G.; Choti, M.A.; Fichtinger, G.; Taylor, R.H. A dual-armed robotic system for intraoperative ultrasound guided hepatic ablative therapy: A prospective study. In Proceedings of the IEEE International Conference on Robotics and Automation, ICRA'04, New Orleans, LA, USA, 26 April–1 May 2004.

14. Hong, J.; Dohi, T.; Hashizume, M.; Konishi, K.; Hata, N. An ultrasound-driven needle-insertion robot for percutaneous cholecystostomy. *Phys. Med. Biol.* **2004**, *49*, 441–455. [CrossRef] [PubMed]
15. Kettenbach, J.; Kronreif, G.; Figl, M.; Fürst, M.; Birkfellner, W.; Hanel, R.; Bergmann, H. Robot-assisted biopsy using ultrasound guidance: Initial results from in vitro tests. *Eur. Radiol.* **2005**, *15*, 765. [CrossRef] [PubMed]
16. Bassan, H.; Hayes, T.; Patel, R.V.; Moallem, M. A Novel Manipulator for 3D Ultrasound Guided Percutaneous Needle Insertion. In Proceedings of the IEEE International Conference on Robotics & Automation, Rome, Italy, 10–14 April 2007.
17. Ghodoussi, M. Robotic surgery-the transatlantic case. In Proceedings of the IEEE International Conference on Robotics & Automation, Washington, DC, USA, 11–15 May 2002.
18. Marescaux, J.; Rubino, F. The ZEUS robotic system: Experimental and clinical applications. *Surg. Clin. N. Am.* **2003**, *83*, 1305–1315. [CrossRef]
19. Guthart, G.S.; Salisbury, J.K. The IntuitiveTM telesurgery system: Overview and application. In Proceedings of the IEEE International Conference on Robotics & Automation, San Francisco, CA, USA, 24–28 April 2000; pp. 618–621.
20. Goonewardene, S.S.; Catterwell, R.; Brown, M.; Challacombe, B. Robotic surgery with the Da Vinci Xi: Simultaneous upper and lower tract surgery. *J. Robot. Surg.* **2016**, *11*, 373–374. [CrossRef] [PubMed]
21. Hannaford, B.; Rosen, J.; Friedman, D.W.; King, H.; Roan, P.; Cheng, L.; Glozman, D.; Ma, J.; Kosari, S.N.; White, L. Raven-II: An Open Platform for Surgical Robotics Research. *IEEE Trans. Biomed. Eng.* **2013**, *60*, 954–959. [CrossRef] [PubMed]
22. Konietschke, R.; Hagn, U.; Nickl, M.; Jorg, S.; Hirzinger, G. The DLR MiroSurge—A robotic system for surgery. In Proceedings of the 2009 IEEE International Conference on Robotics and Automation, ICRA'09, Kobe, Japan, 12–17 May 2009.
23. Hagn, U.; Konietschke, R.; Tobergte, A.; Nickl, M.; Jörg, S.; Kübler, B.; Passig, G.; Gröger, M.; Fröhlich, F.; Seibold, U.; et al. DLR MiroSurge: A versatile system for research in endoscopic telesurgery. *Int. J. Comput. Assist. Radiol. Surg.* **2010**, *5*, 183–193. [CrossRef] [PubMed]
24. Aksungur, S. Remote Center of Motion (RCM) Mechanisms for Surgical Operations. *Int. J. Appl. Math. Electron. Comput.* **2015**, *3*, 119. [CrossRef]
25. Newman, W.S. Stability and Performance Limits of Interaction Controllers. *J. Dyn. Syst. Meas. Control* **1992**, *114*, 563–570. [CrossRef]
26. Kazerooni, H.; Snyder, T.J. Case Study on Haptic Devices: Human-Induced Instability in Powered Hand Controllers. *J. Guid. Control. Dyn.* **1995**, *18*, 108–113. [CrossRef]

electronics

Review

A Review on the Rehabilitation Exoskeletons for the Lower Limbs of the Elderly and the Disabled

Tao Wang [1,†], Bin Zhang [2,†], Chenhao Liu [2], Tao Liu [2], Yi Han [3], Shuoyu Wang [3], João P. Ferreira [4], Wei Dong [5,*] and Xiufeng Zhang [6,*]

1 Nanjing Lishui High-Tech Industry Investment Co., Ltd., 288 Qinhuai Avenue, Lishui District, Nanjing 211200, China; twang@zlznzz.com
2 State Key Laboratory of Fluid Power and Mechatronic Systems, School of Mechanical Engineering, Zhejiang University, Hangzhou 310027, China; zjuzhangbin@zju.edu.cn (B.Z.); 12125056@zju.edu.cn (C.L.); liutao@zju.edu.cn (T.L.)
3 Department of Intelligent Mechanical Systems Engineering, Kochi University of Technology, 185 Miyanokuchi, Tosayamada-Cho, Kami 782-8502, Japan; hanyi1719@126.com (Y.H.); wang.shuoyu@kochi-tech.ac.jp (S.W.)
4 Institute of Superior of Engineering of Coimbra, Quinta da Nora, 3030-199 Coimbra, Portugal; ferreira@mail.isec.pt
5 School of Mechatronics Engineering, Harbin Institute of Technology, Harbin 150001, China
6 Key Laboratory of Rehabilitation Technical Aids Technology and System of the Ministry of Civil Affairs, National Research Center for Rehabilitation Technical Aids, Beijing 100176, China
* Correspondence: dongwei@hit.edu.cn (W.D.); zhangxiufeng@hit.edu.cn (X.Z.); Tel.: +86-186-8689-3076 (X.Z.)
† These authors contributed equally to this work.

Abstract: Research on the lower limb exoskeleton for rehabilitation have developed rapidly to meet the need of the aging population. The rehabilitation exoskeleton system is a wearable man–machine integrated mechanical device. In recent years, the vigorous development of exoskeletal technology has brought new ideas to the rehabilitation and medical treatment of patients with motion dysfunction, which is expected to help such people complete their daily physiological activities or even reshape their motion function. The rehabilitation exoskeletons conduct assistance based on detecting intention, control algorithm, and high-performance actuators. In this paper, we review rehabilitation exoskeletons from the aspects of the overall design, driving unit, intention perception, compliant control, and efficiency validation. We discussed the complexity and coupling of the man–machine integration system, and we hope to provide a guideline when designing a rehabilitation exoskeleton system for the lower limbs of elderly and disabled patients.

Keywords: exoskeleton; intention perception; rehabilitation; medical treatment; efficiency evaluation

Citation: Wang, T.; Zhang, B.; Liu, C.; Liu, T.; Han, Y.; Wang, S.; Ferreira, J.P.; Dong, W.; Zhang, X. A Review on the Rehabilitation Exoskeletons for the Lower Limbs of the Elderly and the Disabled. *Electronics* **2022**, *11*, 388. https://doi.org/10.3390/electronics11030388

Academic Editor: Enzo Pasquale Scilingo

Received: 11 November 2021
Accepted: 19 January 2022
Published: 27 January 2022

1. Introduction

At present, there are a large number of people in China who suffer various movement dysfunctions at different levels caused by stroke, spinal cord injury, and aging. It causes a dual physical and mental impact on patients themselves and brings a heavy medical burden to society and family [1]. Statistics from the Disabled Persons' Federation of China showed that the total number of disabled people had exceeded 85 million, accounting for approximately 6% of the national population, among which about 24 million people were suffering from physical disability [2], which requires care. Stroke is the deadly diseasewith a high mortality and disability rate, which has shown explosive recent growth; 85% of stroke patients lose their walking ability [3]. Moreover, trauma and degeneration are the main causes of paralysis accompanied by motion and sensory dysfunction. There are 2.5 million people worldwide who have suffered from this disease, and the number has increased to 130 thousand annually [4]. The seventh national census of China shows that the elderly above 60 account for 18.7% of the population, while the latest forecast of United Nations population data shows that the elderly population in China will approach

470 million, accounting for more than 30% of the total population in 2025. This means that China will become the country with the highest degree of an aging population in the world, bringing severe challenges to the elderly care service system [5].

Elderly or disabled people with lower limb motor dysfunction, who stay in bed or sit for a long time, will gradually develop a series of complications, such as pressure ulcers, muscle atrophy, organ dysfunction, edema, or osteoporosis, which will further worsen the health condition [6]. In order to improve the quality of life of such people, there is already effective auxiliary walking equipment, such as crutches and wheelchairs, but such devices are unintelligent and inconvenient to use [7] for the people who have lost their moving ability. On the other hand, appropriate medical treatment is also an essential procedure, and rehabilitation doctors need to formulate detailed treatment plans according to the patient's condition and complete regular procedures of body exercise training [8]. However, facing such a large number of patients with movement dysfunction, the relevant professional technicians will be in short supply with the aging population. The recovery of patients depends directly on the professional quality and clinical experience of the technicians. The rehabilitation training process will also seriously consume the physical energy of the physician, which is not conducive to improving efficiency and saving costs. To meet the increasing demand, which is combining the professional rehabilitation procedures with daily assistance, exoskeleton technology has drawn increasing attention to achieve intelligent training and evaluation.

The vigorous development of exoskeletal technology has brought new ideas regarding the rehabilitation and medical treatment of patients with motor dysfunction, which is expected to help such people complete their daily physiological activities or even reshape their motion function [9]. To fully understand the efforts on rehabilitation exoskeletons, this paper reviews the published works on rehabilitation exoskeletons from 2003 to 2021 in the Web of Science database [10]. When the papers were reviewed, keywords "rehabilitation exoskeleton" and "lower limb exoskeleton" were combined with "for the disabled" or "for the elderly" to collect the published literature. The keywords search generated more than 136 journal and conference papers related to rehabilitation exoskeletons. Papers not related to the research topic, repetitive articles, and articles related to walking aids for the blind and children's rehabilitation aids were excluded. We selected 96 papers to review the exoskeletons from the point of manufacturing the rehabilitation exoskeletons. Based on their contents, the key points of these papers can be categorized into five aspects: ergonomic design, actuation, perception, control, and validation methods.

The exoskeleton systems were identified as wearable man–machine devices made through anthropomorphic design, providing active assistance to the users according to their motion intention. This is one of the most promising potential technical studies to deal with the problems of disabled care and elderly assistance and rehabilitation [9]. The design of rehabilitation exoskeletons should match the ergonomic principles to guarantee that the system can correspond with the distribution of human joints. To achieve rehabilitation procedures, the structure of the system should be specified, including sensors, actuators, and controllers. Based on our surveyed literature, most exoskeletons were actuated by motors and some artificial actuators such as pneumatic muscles (PMs) [11] and shape memory alloy (SMA) [12] were reviewed. The control methods determined the performance of the exoskeletons conducting assistance and rehabilitation. The control algorithms were set to maintain human–machine interaction and send the precise control commands to drive the actuators to perform corresponding auxiliary actions. Most interaction methods were achieved by feedback control based on detecting the information of the exoskeletons, the users, or the man–machine coupling system. Many behavioural and physiological sensors were introduced into the rehabilitation system to represent the status of the man–machine system, where the behavioural and physiological sensors describe the kinematic features (such as joint angles, velocity, acceleration, etc.) and human physical status (such as heart rate, Electromyogram (EMG), electroencephalogram (EEG), etc.). Whether it is helping patients with daily physiological activities such as walking or performing regular rehabilitation training in accordance with the treatment plan, the exoskeleton system could easily and effectively complete the rehabilitation goals. Finally, it is necessary to evaluate the effectiveness of the exoskeleton systems and the performance of the assistance and rehabilitation.

Developed countries represented by the United States, Switzerland, and Canada started to explore the design and application of rehabilitation medical exoskeleton robots very early [13]. So far, a series of commercial products have been developed to meet various needs, which has greatly promoted the research process of exoskeleton technology in helping the elderly and the disabled. During the 11th Five-Year Plan of China, according to the growing demand for domestic social development (especially the elderly and the disabled), rehabilitation robots started as a key research project. In the 12th Five-Year Plan, a further concept was proposed regarding rehabilitation. The 13th Five-Year Plan carried out a continuous special research plan for rehabilitation robots [14]. At present, both universities and institutes have been committed to the research, and a large number of rehabilitation exoskeletons have also emerged [15].

The ergonomic design determined the matching performance of the exoskeletons, which plays a fundamental role to achieve assistance and rehabilitation. The actuators, sensors, and controllers are the basic elements to conduct motion, perception, and control, which are the essential parts of the rehabilitation exoskeletons. The validation methods were designed to confirm the effectiveness of the rehabilitation procedure and the exoskeletons. In this paper, we reviewed the rehabilitation exoskeletons from the aspects of ergonomic design, actuation, perception, control, and validation. We discussed the advantages and limitations of the man–machine interaction systems and stated our considerations of designing and developing the rehabilitation exoskeletons in the future. We hope this paper can provide an overall guideline to design a rehabilitation exoskeleton system. The contributions of this paper include: (1) as a paper dedicated to reviewing the rehabilitation exoskeletons, five aspects are listed to summarize state of the art technologies; (2) the advantages and limitations for every aspect are proposed; (3) the challenges of the ergonomic design, sensor-based motion perception, actuation, and control were discussed. We hope this paper could provide a guideline when designing a rehabilitation exoskeleton system for the lower limbs of the elderly and disabled.

2. Design of the Rehabilitation Exoskeletons

Due to the differences in the degree and location of the loss of motor function of the wearer, the structural forms of the exoskeleton helping the elderly can be designed according to the different rehabilitation goals. The unpowered exoskeletons, which do not contain any powered elements (such as a battery, electric motor, etc.), provided a rich experience for the ergonomic design because the matching performance determined the distribution and transmission of the force [16]. There were already several representatives of passive exoskeletons, such as OX, UPRISE, Niudi, and FORTIS [16]. The OX was designed by the Australian Government Department of Defense, and it can transfer two-thirds of the pressure borne by a soldiers' shoulders, spine, and legs to the ground. Mawashi Co. (Quebec, Canada) developed UPRISE transferring 50–80% of the pressure borne by a soldiers' shoulders to the ground without interfering with normal motion. UPRISE is constructed by using high-strength titanium alloy. Niudi Co. (Chongqing, China), LTD from China proposed a modularized UE that can withstand 70 kg but weighs only 6 kg. The FORTIS is designed by Lockheed Martin Co. (Bethesda, MD, USA) to help workers carry heavy tools. The unpowered exoskeletons have great potential in the military, industry, rescue fields, etc. The passive exoskeletons were well-bionic designed and constructed by dexterous structure. The ergonomic design of the passive exoskeleton could be referred to as the design of rehabilitation exoskeletons.

The rehabilitation exoskeletons were designed for the elderly and the disabled who have been suffering from moving dysfunction. Compared with the power-enhanced exoskeleton worn for people with normal mobility, the safety and stability of the man–

machine interaction process must be guaranteed through the special design of the system itself in structure. To achieve this basic requirement, the overall scheme of the existed systems was divided into a platform-based exoskeleton, crutches-based exoskeleton, and self-balanced exoskeleton. The Swiss Locomat [17] and American ALEX III [18] belong to the first type of rehabilitation exoskeleton. The users were equipped with lower limb exoskeletons under the protection of the weight support structure and completed the walking process on the treadmill, which can safely achieve the intensive training of lower limb muscle strength, as shown in Figure 1c,d. The second type of system combines crutches with the exoskeleton. The Israeli ReWalk series [19] introduced the users' upper limbs to maintain stability by using crutches for patients with lower motion dysfunction, as shown in Figure 1b. The mode switching buttons were set in the crutches helping the users adjust movement modes such as tuning walking speed, navigation, and interaction. The third category uses the balance control algorithm to automatically adjust the movement posture of the human–machine system, which can operate normally without structural assistance, such as the New Zealand Rex [18].

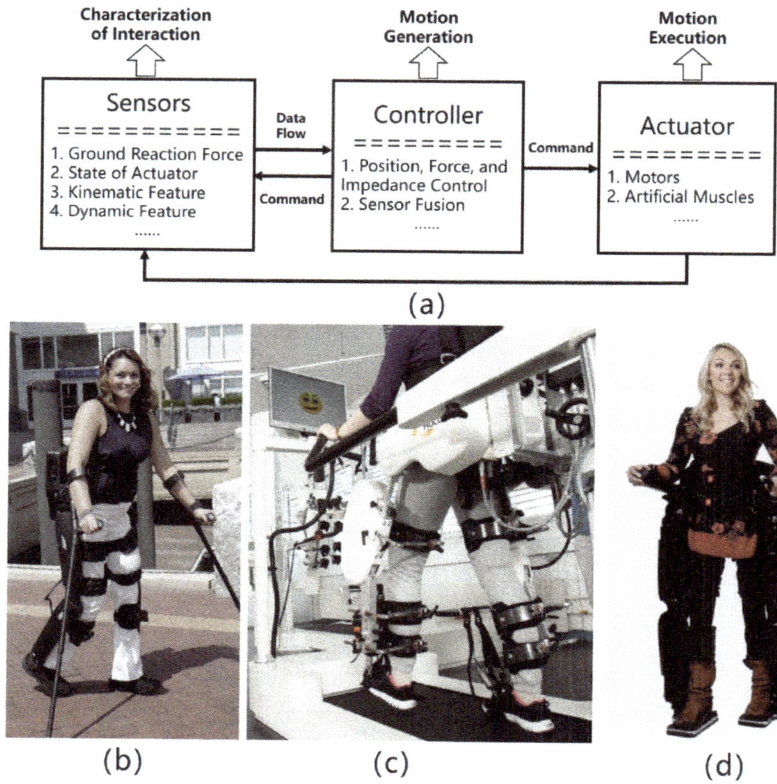

Figure 1. Overall exoskeleton plan for helping the elderly and the disabled. (**a**) Schematic diagram to describe the structure of exoskeletons. (**b**) Based on crutches, reprinted from ref. [19]; (**c**) Based on the platform, reprinted from ref. [17]; (**d**) Based on the self-balance design, reprinted from ref. [18].

In terms of active freedom configuration, the existing exoskeletons also showed different characteristics, and the summary of the existing exoskeletons is shown in Table 1. The working form of the rehabilitation exoskeletons were divided into treadmill based and over-ground. The treadmill-based exoskeletons constructed a specific trajectory in space, and the patients' legs were constrained [17]. The over-ground exoskeletons usually allowed the patients to walk on the ground [18,19]. Motors were still the common actuators to drive the motion of the exoskeleton, which were easily controlled based on the developed algorithms. The devices were designed based on the level of losing mobility. Most of the exoskeletons were targeted to the hip joints and could assist the patient to walk based on balance control. Some exoskeletons introduced crutches to avoid a tumble. However, there is still a lacking of standard validation methods to confirm the effectiveness of the exoskeletons. Some systems drive a single joint such as the knee and ankle, and the wearer often participates in a fixed posture to complete daily physical activities such as walking. In order to enable the exoskeleton to assist patients with basic gait and active hip and knee, such as Indego of Parker Hannifin Corporation [20–23], ROBIN of Korea Industrial Technology Institute of Parker Hannifin [24,25] and MINDWALKER of Delft Polytechnic University in the Netherlands. In addition [26], a few systems set multiple active joint mobilities to achieve self-balance or motor flexibility. For example, Rex of Rex Bionics, New Zealand and ATLAS 2020 [27], developed by the Spanish National Research Council, have 10 active degrees of freedom.

Table 1. Summary of existing exoskeletons and their related technical details (treadmill-based exoskeletons: The man–machine system operates on the treadmill. Over-ground: The man-machine system can operate on the ground).

Devices	Working Form	Actuator	Control Strategies	Target Parts of Human	Keywords of Validation Methods
Lokomat [17]	Treadmill based exoskeleton	Motor	Position and impedance	Hip and knee	EM activity
ALEX III [18]	Over-ground	Motor	Balance and impedance	Hip, knee and ankle	Reshape walking ability
ReWalk [19]	Over-ground	Motor	Force and impedance	Hip and knee	Walking assistance
Indego [23]	Over-ground	Motor	Position and force control	Hip and knee	Walking assistance
Mindwalker [26]	Over-ground	SEA-motor	Electroencephalogram (EEG)-based position control	Hip, knee and ankle	Reshape walking ability
BLEEX [28]	Over-ground	Hydraulic Cylinders	Position and force control	Hip, knee and ankle	Reshape walking ability
Ankle-foot Exoskeleton [29]	Over-ground	Pneumatic Muscle	Position and force control	Ankle	Reduction in Metabolic (21%)
LOPES [30]	Treadmill based exoskeleton	Bowden Cable-based Series Elastic Actuator	Torque	Hip, knee and ankle	Improving the control compliance
Rex [31]	Over-ground	Motor	Balance and force control	Hip, knee and ankle	Reshape walking ability
HAL [32]	Over-ground	Motor	EMG-based force control	Hip, knee and ankle	Walking assistance

The wearable form of the rehabilitation exoskeletons guarantees the exoskeletons can assist in an ergonomic way. The exoskeletons must be designed based on the distribution of the humanoid characteristics such as muscle distribution, tendon-based transmission, and skeleton-based support. The position of the actuators should be placed along with human joints. The passive exoskeletons applied in military and industry fields showed great development in ergonomic design. Unlike the unpowered exoskeletons, the external energy should be introduced into the man-machine system because the rehabilitation exoskeletons are targeted to reshape or maintain the mobility of the people with moving dysfunction. Therefore, the rehabilitation exoskeletons should be powered exoskeletons, including actuators, sensors, and control methods. In our surveyed literature, the motors were the most traditional actuators used in rehabilitation systems, which have been developed for decades. Some novel actuators were introduced into the rehabilitation exoskeletons, such as PM and SMA, inspired by their bionic characteristics.

3. Actuation

The joint actuators belong to the execution part of the exoskeleton movement, delivering the desired power to achieve the auxiliary movement. The moving performance was determined by key characteristics such as the power effect, composition shape, and response speed of the actuators. In terms of the current technology of the actuators of exoskeletons for the elderly and the disabled, it can be divided into three aspects motor drive, air pressure drive, and functional electric stimulation based on surveyed papers.

Motor-driven exoskeleton drives the system joints by a straight or rotating motor. Most rehabilitation training equipment with compact structure and fast response was driven by electric motors [33–37], as illustrated in Figure 2b. Meanwhile, the structure of the rehabilitation exoskeletons was bulky because of the big batteries and motors. Galle et al. [29] proposed a powered ankle–foot exoskeleton to reduce the metabolic cost which is driven by pneumatic artificial muscle, as shown in Figure 2b. Timing control was optimized and implemented, and the experimental results showed a 21% reduction in metabolic cost. However, prolonged rehabilitation training causes muscle fatigue, such as in the US Vanderbilt system [38] (Figure 2c). Similar to the PM-actuated exoskeletons, the hydraulic exoskeletons depended on the pressure supplies [28]. The electrohydraulic actuator included a motor, gear pump, and antagonistic installed cylinders [39]. The cylinders were controlled by servo valves and powered by combustion engines or electric motors.

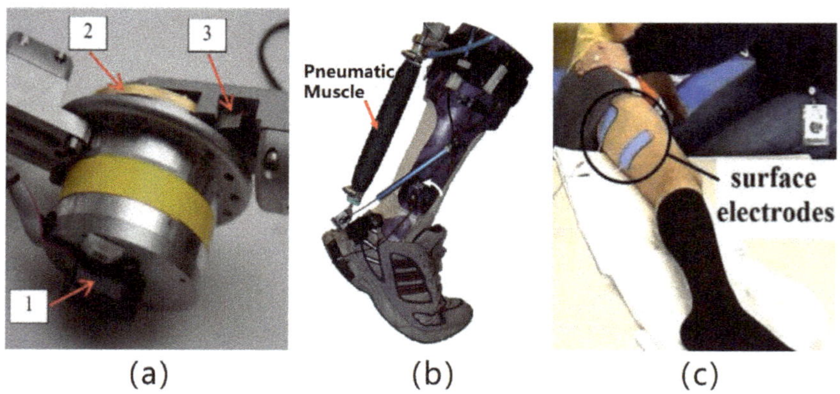

Figure 2. Exoskeleton's actuators. (**a**) Motor drive, adapted from ref. [33]; (**b**) Air pressure drive reprinted from ref. [29]; (**c**) Self muscle drive, reprinted from ref. [40].

In addition, new driving solutions for joint assistance are also being explored and developed. Chinese Academy of Sciences applies magnetic rheological actuator to robot rehabilitation system, which can act as a brake or clutch according to the working state of the system (human activity mode/manual activation mode) for energy consumption saving [41]. The University of Carlos III introduced the design, test, and analysis of SMA drives under various configurations and explained the feasibility of the method in soft robots and light exoskeletons [42].

The electric motors were still the most used actuators for the rehabilitation exoskeletons. Generally, the motors increase the torque by decreasing the rotation speed by the reduction boxes. That is how the electric motors were limited by requiring transmission elements to convert their high-speed, low-torque output features to low-speed and high-torque features [30]. The rehabilitation exoskeletons could provide a powerful force to lift users' body and keep balance. However, the rehabilitation exoskeletons were of large volume because of the motors, reducers, and batteries which made the system bulky and nonflexible. Therefore, there was a tradeoff between the flexible system and powerful output force. For the rehabilitation process, the movements of rehabilitation were achieved slowly to guarantee safety during training. It implied that the motors were suited to assist people with slow speed, which was appropriate for the disabled and the elderly.

The PMs were studied for decades from modelling, design, and control [43,44]. The PMs were compliant actuators mimicking human muscles, which were lightweight, had a high power-to-weight ratio, and were compliant [45]. However, there was an inevitable hysteresis between the inflation and deflation process. The hydraulic actuators were selected to drive exoskeletons because of their high-specific power [39], high-force output [28], wide control bandwidth, smooth actuation [46] etc. However, a significant limitation of the pneumatic and hydraulic actuators was the dependence on non-portable pressure supplies. Both the pumps and air compressors were too heavy and large to carry. Their applications were limited to the platform-based rehabilitation fields with no or low portability.

The other novel actuators were proposed and applied to the rehabilitation system. The SMA-actuated soft exoskeletons achieved finger motions. Due to the limited scale of the output force, novel actuators were suitable for precise rehabilitation with a small motion range. There is still a long road to transfer the novel actuators from lab to prototype.

4. Motion Intention Perception

Essentially, every person has a unique gait pattern, and it is inaccurate to set a fixed gait pattern [38]. Therefore, it is necessary to percept human motion to provide feedback to the exoskeleton system. Human motion intention recognition is involved in obtaining man–machine interaction and state information, data fusion algorithm to achieve the purpose of human motion, or even predict the human motion based on multiply sensors.

According to the difference in intention information acquisition mode, it can be divided into three intention perception methods based on motion signal, biological signal and mixed man–machine hybrid signal [38]. The motion signals described kinematic and dynamic characteristics such as motion acceleration, walking speed, joint angles etc., which can be measured by the gyros or accelerators. The biological signals represented the physiological information such as muscle fatigue, neural activation, etc., which can be measured by EMG, EEG, etc. The man–machine hybrid signals described the interaction forces.

The first method had been widely used in typical exoskeletal systems such as the LOPES [47] American eLEGs [48] through conventional sensors distributed in man–machine connections or system motor parts, such as an inertial measurement unit mounted on the limb, a contact force sensor in insoles. The second method could effectively reduce the hysteresis of intention perception based on extracting and processing EEG or myography data. For example, the University of Houston used non-invasive brain-computer interface technology to control the Neuro Rex exoskeleton in real-time, explain the wearer's intention through EEG signals and finally achieve the goal of assisting paraplegic patients to complete walking independently [31]. The HAL exoskeleton developed by the University of Tsukuba in Japan can detect and obtain electrical signals generated by muscle movement on the skin surface, which was processed as input to the system control [32]. The third method comprehensively considers the advantages and disadvantages of the above two methods and establishes a man–machine sensor network to comprehensively utilize biological and motion signals. The high reliability and low latency of intention perception have now become one of the focuses of research on exoskeleton technology.

A single sensing data can only reflect limited information of the man-machine interaction process. It requires the fusion processing of multimodal information of different sources, different levels and different manifestations to ensure the speed and accuracy of intention perception. For data fusion algorithms, according to the difference of data processing levels, it can be divided into data level, feature level, and decision level fusion. More common methods include D-S evidence theory, artificial neural network, adaptive weighted average, Bayesian estimation, fuzzy set theory, etc. For example, Northwestern Industrial University has achieved data fusion of EEG and EMG through D-S (Dempster/Shafer) evidence theory and backpropagation neural network [49]. The integration of the data collected by the three sensors improves the recognition accuracy of various motion modes.

Human motion information has been drawing researchers' attention from the aspects of the portable and wearable form, high accuracy, and low latency. Fortunately, the motion sensors can be manufactured into tiny types based on the current semi-conductor development. However, the accuracy of the motion recognition was still limited. The most popular physiological information was represented by EEG and EMG, depicting the command from the brain and execution from muscles [31,32]. Moreover, there were no non-invasive sensors with a high signal–noise ratio measuring the intention information [50]. The invasive sensors showed great potential in improving signal quality, but it is still unclear whether the invasive sensors interfere with normal human motion. MINDWALKER combined the EMG and EEG to extract the motion features to improve the accuracy [26,51,52].

Besides, motion intention could be obtained by the kinematics parameters. Indego [23] introduced inertia measurement units (IMU) to recognize gait patterns. Then, they tuned the center of pressure to track the desired movements. Peruzzi et al. [53] implemented gait evaluation based on multiple IMUs. Integrating was conducted to obtain the velocity and displacement. However, the IMUs-based motion detection would introduce inevitable errors when conducting integration. Wang et al. [54,55] implemented zero velocity state detection by judging the difference between the forward acceleration measured by the accelerator and the forward acceleration derived by the differential of angular velocity measured by gyros.

The most efficient sensors with a high signal–noise ratio and low latency represented the motion behind the human, which introduced inevitable latency of the man–machine system. Motion prediction based on kinematic and dynamic information could solve the problem essentially. It is still feasible to use the current sensors to conduct perception during rehabilitation because the rehabilitation process was usually slow-speed, and the latency of the sensors was acceptable in the slow process of rehabilitation.

5. Control Methods

The exoskeleton compliant control strategy is based on the result of intention perception. Controlling the joints to assist the human body in accordance with the wearer's motion intention or the rehabilitation physician's plan was key to ensure that the system generated the desired motion and produced auxiliary effects. Li et al. sorted the control strategies for lower limb rehabilitation exoskeletons with eight categories [56]. At present, common control algorithms for helping the elderly and the disabled mainly include gait trajectory planning, impedance/guide control, biological signal-based control algorithm, etc.

The trajectory planning control was usually used to perform movement tasks when humans lost the motor ability completely and conducted passive rehabilitation at the early stage of damage [57]. In the gait trajectory planning algorithm, the exoskeleton joints were controlled to produce periodic motions according to the pre-designed path, which simulated the gait of humans completing daily activities achieving the coordination of human and machine actions. The reference movement trajectory can be set in a variety of ways. Firstly, the gait trajectory playback strategy that directly uses healthy human gait parameters was widely used in rehabilitation medical exoskeleton. Systems such as WPAL [58,59] in Japan and Mina [60,61] in the United States can collect gait data of normal wearers and reproduce the movement process when assisting patients. Secondly, a mathematical model-based gait trajectory generation method can also be employed to calculate the required motion parameters through related theories. Rex from New Zealand generated a gait based on the ZMP model so that the system had self-balancing ability. ATLAS in Spain planned the corresponding motion trajectory based on the inverted pendulum model and calculated the key parameters such as step length and step height required to complete the desired gait [62].

Furthermore, to activate the muscle participation in the rehabilitation process and achieve compliant interaction torque and voluntary muscle torque, the impedance/admittance model was established to realize the mixed control of the force and position of the joint motion. The impedance control could provide assistance that is proportional to the difference between the human limb and the given trajectory [63,64]. McGill University in Canada designed a new adaptive impedance control strategy, which combined backstepping control, time delay estimation, and interference observer, improving the effect of passive assisted rehabilitation training [65]. The University of Twente in the Netherlands had designed and verified the admittance controller on the LOPES II exoskeleton, which could generate the gait trajectory of the impaired based on the healthy leg [66].

The most classic biology-based control methods were based on EMG and EEG [57]. Both of them control the exoskeletons based on the direct motion intention [67,68]. The control algorithm based on biological information analyzed the control input required for the movement of the exoskeleton on the basis of collecting and analyzing the bioelectric signal of the human body. The Italian Institute of Technology combined the Hill muscle model and EMG signals to estimate the driving torque required for the knee joint to complete the exercise in real-time [69]. The University of Michigan in the United States used an adaptive gain proportional EMG controller for ankle joint assistance, relying on dynamic gain to map the wearer's muscle activity to actuation control signals. The experimental trials showed that the system significantly enhanced ankle strength and reduced metabolic consumption [70]. EEG-based control is manipulating the exoskeletons based on the brain–computer interface (BCI). Evidence has shown that a patient with tetraplegia could be able to control an exoskeleton by using BCI [71]. Vouga et al. [72] enabled a monkey wearing an exoskeleton to track the cursor on a screen based on continuous regression methods. EEG parsing has been the hot research for decades, and they usually resorted to a black box model. EEG has shown great potential in controlling rehabilitation exoskeletons [73].

The interaction control theory has been developed for a long time, and it has been rather mature for the rehabilitation exoskeletons during one rehabilitation process. There was no need to demand the real-time performance of the control methods because the rehabilitation process was usually reduced to a slow process [9]. However, there was lacking rehabilitation strategy control which usually was made by the doctors. The rehabilitation strategy is to set the rehabilitation control with different groups of parameters related to users' status. The optimization methods should be implemented to select the best parameters based on the judgement of the rehabilitation stage.

6. Validation of the Rehabilitation Exoskeleton

Quantitative evaluation of the auxiliary effect of the exoskeleton was necessary to test and optimize the system. The evaluation indicators need to be determined according to the requirements of rehabilitation training. The efficiency evaluation of rehabilitation exoskeletons can be divided into task-based evaluation, kinematic and kinetics evaluation, and interaction evaluation. In task-based evaluation, walking speed and walking distance are direct indicators of the training results, and the corresponding data can be obtained through the 5 min walking test (5MWT) [9], 6 min walking test (6MWT), 10 m walking test (10MWT) [74], and timed up and go (TUG). Besides, endurance [75], versatility [76], and max speed [77] were used to evaluate the performance of the rehabilitation exoskeletons under specific tasks.

The validation methods for rehabilitation exoskeletons focuses on sensors used for biomechanics and energetics measurements. In general, kinematic and dynamic measurement was used to evaluate the flexibility of the rehabilitation exoskeletons and to predict energy expenditure indirectly. The most popular kinematic parameters were used to validate the exoskeletons were joint angular trajectories, range of motion (ROM), speed, and COM position [9]. Joint torque output [78], peak power [79], and maximal torque [80] were used to test the validation of the proposed exoskeletons. Metabolic cost measurement represented how much energy was saved by the rehabilitation exoskeletons. The EMG signal represented the activity of the muscle, which can be used to describe the fatigue of the muscles [74]. Kinematic parameters describe the changes of displacement in linear or angular position and their derivatives, such as linear velocity, linear acceleration, angular velocity, and angular acceleration. They can be obtained by the camera-based recognizing and wearable recognizing system. Kinetics aims to study forces that affect human motion. These forces can change the linear or angular motions. Force data can be obtained directly by using force and torque sensors. GRF is the representation of the human body's impact on the ground measured by force plates, which can be used to analyze the force provided by exoskeletons. The energy for human motions comes from the chemical energy by digesting food, and it flows in three directions: entropy, maintenance, and muscle energy. The energy of metabolic cost is part of muscle energy, and it can be measured indirectly by recording respiratory flow, respiratory flow rate, heart rate, muscle activity, etc.

Comfort reflected the subjective sense of patient interaction with the device. Generally, we collected the feedback about comfort based on the questionnaire. The strategies of online detection of physiological data such as heart rate and oxygen consumption can also be used to intuitively reflect the physiological state of the human body. Besides, the ergonomic parameters determined the matching performance when humans wore the exoskeletons, which represented the effectiveness of the design [9]. However, the current evaluation methods were good at confirming the effectiveness of the exoskeletons, but they failed to confirm the effectiveness of the rehabilitation process. Clinically, doctors adjust the rehabilitation process based on the score-based evaluation methods, which is currently the golden standard [81]. The multi-information collected by the sensors should be introduced into the evaluation process to confirm the rehabilitation stage.

7. Discussion and Conclusions

At present, countries all over the world have achieved considerable research results concerningexoskeleton technology for rehabilitation medical treatment [9]. Prototypes and commercial products with different forms and functions are emerging, bringing new ways to help the rehabilitation of the elderly and the disabled. However, in terms of the current technical conditions, it is still facing serious challenges to realize the comprehensive promotion of the exoskeleton in the social rehabilitation service system.

1. The matching performance between the exoskeletons and the human body is significantly low. The exoskeleton system with more active degrees of freedom has better flexibility, but complex structural composition and hardware lines affect the overall performance of the system. The most popular ergonomic indicators were human-robot relative position, interface displacements, anthropometric database percentiles, and adaptability to different height ranges [82]. The joints of the human body and exoskeleton have obvious rotation centers misalignment during movement, which will reduce man-machine interaction. The coupling system may be deformed and misaligned during the interaction, which may reduce the power-assisted effect. Problems similar to the above are widespread in existing systems and severely disrupt ergonomics. To make the wearable exoskeletons more comfort, the unpowered exoskeletons inspired the novel design of rehabilitation exoskeletons. The materials and manufacturing methods for lower limb exoskeletons are important because they guarantee a safe and ergonomically comfortable interface with the human [83,84].
2. Sensor-based motion feedback is the basis of exoskeleton controlling and rehabilitation [56]. The joint angle and interaction torque were the frequently used feedback in most studies. The joint angle can be used to describe the difference between the given trajectory and output and calculate the force by joint angle deviation [85], the impedance by derivation of the ankle joint [86]. The interaction torque was usually used to generate real-time trajectory [87] and provided reference to correct the trajectory [88]. The recognition and prediction of human movement intention are not accurate enough. The motion intention is estimated by the mechanical signal obtained by the sensor; although the result is reliable and stable, there is a large hysteresis which is generated by the signal conversion and decoding process [89]. The human intention based on biological signal analysis has good timeliness, but the data is unstable, and

the use process is cumbersome. Therefore, it is necessary to set up a steady communication pipeline between the rehabilitation exoskeletons and humans. Moreover, there were many kinds of existing data fusion algorithms such as radial basis function neural networks [90], convolutional neural networks [91], musculoskeletal model [92], etc., but motion mode recognition is often difficult to meet the requirements of safety and reliability accuracy. Previous work has proposed machine learning-based predictors based on EMG, kinetics, and kinematics to estimate the desired motion intention. More recently, several researchers have explored using teleceptive sensing of terrain to improve the prediction of desired locomotion [93]. The above difficulties are urgent problems to be solved in human intention recognition and prediction. It is feasible to combine the physical model of human motion with the motion data to achieve fast and stable intention perception [93]. The body domain network is designed to obtain the information of human body movement, physical parameters and state.

3. Lightweight and high power/weight ratios of driving units are difficult to achieve. Existing exoskeletons' actuators are often with lower power/weight ratios, and they have large volumes and mass, such as Lokomat [63] and ALEX [18], resulting in large and bloated overall structural forms. Hydraulic [39] and pneumatic actuators [11] improved the power/weight ratio but introduced non-portable pressure supplies and control difficulties. The series elastic actuators combined the performance of easy to control and the compliant features [67]. However, it is necessary to explore more effective driving forms as well as innovative design and optimization methods of high-power density driving units. A permanent magnet servo motor should be designed by using conservative optimization design methods to realize the lightweight of permanent magnet servo motor [94]. The speed closed-loop control strategy and position closed-loop control strategy should be designed to drive the permanent magnet motor to achieve high precision control.
4. The deviation between system motion control and human motion is prominent. The strong autonomy of human motion, the strong coupling of man-machine interaction, and the complexity of the system model have made it difficult for many control algorithms to achieve the goal of man-machine coordination and interaction [56]. The exoskeleton system should meet the needs of the wearer to complete all kinds of basic movements and basic movement transformation. For the passive rehabilitation process, the trajectory-based control is enough to replay the predefined trajectory [60]. But the predefined trajectory is not suitable for different individuals. Introducing human motion intention input into man-machine interaction control was an effective method to achieve more dexterous in assisting human motion [95]. Dynamic control strategy should be implemented in the rehabilitation system, and the stability should be confirmed based on real-time state detection and stability criteria. Dynamic control involved the dynamic modelling of the system, for example, a simple mass-spring-damper model to characterize series elastic actuator [96], actuated dynamic model [97], and hybrid dynamic model [98]. Finally, optimization control methods [99] should be introduced to ensure the reliability and consistency of the rehabilitation.

A number of typical products, such as Lokomat [17], Rewalk [19], HAL [32], etc., have been successfully developed, and application verification has been initially carried out. However, due to the difficulty of lightweight design, weak motion intention identification ability and poor motion control, it is difficult to obtain the qualitative efficiency to improve the existing exoskeleton assistance. It does not have the technical level of system lightweight, accurate identification and smooth motion, which restricts the promotion and application of such exoskeletons. Therefore, it is urgent to study high torque density motor lightweight driving system design theory and method. The multi-mode human movement biological information decoding and transmission mechanism should be revealed [92], and a multi-source body and exoskeleton coordination movement compliant control strategy need to be established. The final goal is to solve the key scientific problems in the engineering application of the exoskeleton robot for the elderly and the disabled and provide the theoretical foundation and technical support for the development of wearable electromechanical systems.

Author Contributions: Conceptualization, W.D. and X.Z.; methodology, T.W. and Y.H.; validation, T.L., S.W. and J.P.F.; writing—original draft preparation, B.Z. and C.L.; writing—review and editing, T.W. and B.Z. All authors have read and agreed to the published version of the manuscript.

Funding: This work was supported in part by the National Natural Science Foundation of China under Grant U1913601, 52175033, and U21A20120, Zhejiang Provincial Natural Science Foundation of China: LZ20E050002, State Key La-boratory of Fluid Power and Mechatronic Systems (Grant No. GZKF-202101), and DongGuan Innovative Research Team Program (2020607202006).

Institutional Review Board Statement: Not applicable.

Informed Consent Statement: Not applicable.

Data Availability Statement: Not applicable.

Acknowledgments: We are grateful to Ning Zhang (Key Laboratory of Rehabilitation Technical Aids Technology and System of the Ministry of Civil Affairs, National Research Center for Rehabilitation Technical Aids, Beijing 100176, China, zhangning@nrcrta.cn) who made profound contributions for the project works.

Conflicts of Interest: The authors declare no conflict of interest.

References

1. Xie, X.-Y.; Lin, G.-Z.; Huang, Q.; Li, C.-B.; Hallett, M.; Voon, V.; Ren, R.-J.; Chen, S.-D.; Wang, G. Opinions and clinical practice of functional movement disorders: A nationwide survey of clinicians in China. *BMC Neurol.* **2021**, *21*, 435. [CrossRef]
2. Xu, S. Development Process and Legislative Suggestions Regarding Disability Welfare Rights in China. *Political Theol.* **2019**, *20*, 411–423. [CrossRef]
3. Li, N.; Yang, J.; Feng, X.; Zhang, J.; Yang, X.; Zhang, Z. A summary of 30 years' research on risk factors of stroke mortality in China. *Chin. J. Behav. Med. Brain Sci.* **2017**, *26*, 765–768.
4. Li, Z.; Singhal, A.; Wang, Y. Stroke Physician Training in China. *Stroke* **2017**, *48*, E338–E340. [CrossRef] [PubMed]
5. Spence, J.D. China Stroke Statistics 2019: A wealth of opportunities for stroke prevention. *Stroke Vasc. Neurol.* **2020**, *5*, 240–241. [CrossRef]
6. Simpson, L.A.; Eng, J.J.; Hsieh, J.T.C.; Wolfe, D.L. The health and life priorities of individuals with spinal cord injury: A systematic review. *J. Neurotrauma* **2012**, *29*, 1548–1555. [CrossRef] [PubMed]
7. Naoaki, T.; Takuya, E.; Yoshihiko, N.; Norihiko, K. Training to Improve the Landing of an Uninjured Leg in Crutch Walk Using AR Technology to Present an Obstacle. *J. Robot. Mechatron.* **2021**, *33*, 1096–1103.
8. Manuli, A.; Maggio, M.G.; Stagnitti, M.C.; Aliberti, R.; Cannavo, A.; Casella, C.; Milardi, D.; Bruschetta, A.; Naro, A.; Calabro, R.S. Is intensive gait training feasible and effective at old age? A retrospective case-control study on the use of Lokomat Free-D in patients with chronic stroke. *J. Clin. Neurosci.* **2021**, *92*, 159–164. [CrossRef]
9. Zhou, X.; Liu, G.; Han, B.; Li, H.; Zhang, L.; Liu, X. Different Prevention and Treatment Strategies for Knee Osteoarthritis (KOA) with Various Lower Limb Exoskeletons—A Comprehensive Review. *Robotica* **2021**, *39*, 1345–1367. [CrossRef]
10. Web of Science. Available online: http://www.isiknowledge.com (accessed on 12 October 2020).
11. Goergen, R.; Valdiero, A.C.; Rasia, L.A.; Oberdorfer, M.; de Souza, J.P.; Goncalves, R.S. Development of a pneumatic exoskeleton robot for lower limb rehabilitation. In Proceedings of the IEEE International Conference on Rehabilitation Robotics (ICORR), Toronto, ON, Canada, 24–28 June 2019; pp. 187–192.
12. Copaci, D.; Flores, A.; Rueda, F.; Alguacil, I.; Blanco, D.; Moreno, L. Wearable Elbow Exoskeleton Actuated with Shape Memory Alloy. In *Converging Clinical and Engineering Research on Neurorehabilitation II*; Springer: Berlin/Heidelberg, Germany, 2019. [CrossRef]
13. Nasiri, R.; Shushtari, M.; Rouhani, H.; Arami, A. Virtual Energy Regulator: A Time-Independent Solution for Control of Lower Limb Exoskeletons. *IEEE Robot. Autom. Lett.* **2021**, *6*, 7699–7705. [CrossRef]
14. Vidal, A.F.P.; Morales, J.Y.R.; Torres, G.O.; Vazquez, F.D.S.; Rojas, A.C.; Mendoza, J.A.B.; Cerda, J.C.R. Soft Exoskeletons: Development, Requirements, and Challenges of the Last Decade. *Actuators* **2021**, *10*, 166. [CrossRef]
15. Hyakutake, K.; Morishita, T.; Saita, K.; Fukuda, H.; Abe, H.; Ogata, T.; Kamada, S. Effect of Robot-assisted Rehabilitation to Botulinum Toxin a Injection for Upper Limb Disability in Patients with Chronic Stroke: A Case Series and Systematic Review. *Neurol. Med.-Chir.* **2021**, *62*, 35–44. [CrossRef] [PubMed]
16. Zhang, B.; Liu, T.; Zhang, B.; Pecht, M.G. Recent Development of Unpowered Exoskeletons for Lower Extremity: A Survey. *IEEE Access* **2021**, *9*, 138042–138056. [CrossRef]
17. Lokomat. Available online: https://www.hocoma.com/media-center/media-images/lokomat (accessed on 10 November 2021).
18. Rex. Available online: https://www.rexbionics.com (accessed on 10 November 2021).

19. Rewalk Personal 6.0 Exoskeleton: More Than Walking. Available online: https://rewalk.com/rewalk-personal-3 (accessed on 10 November 2021).
20. Hartigan, C.; Kandilakis, C.; Dalley, S.; Clausen, M.; Wilson, E.; Morrison, S.; Etheridge, S.; Farris, R. Mobility outcomes following five training sessions with a powered exoskeleton. *Top. Spinal Cord Inj. Rehabil.* **2015**, *21*, 93–99. [CrossRef]
21. Dalley, S.A.; Hartigan, C.; Kandilakis, C.; Farris, R.J. Increased Walking Speed and Speed Control in Exoskeleton Enabled Gait. In Proceedings of the 2018 7th IEEE International Conference on Biomedical Robotics and Biomechatronics (Biorob), Enschede, The Netherlands, 26–29 August 2018; pp. 689–694.
22. Andrew, E.; Michael, G. Supplemental stimulation improves swing phase kinematics during exoskeleton assisted gait of SCI subjects with severe muscle spasticity. *Front. Neurosci.* **2018**, *12*, 374.
23. Juszczak, M.; Gallo, E.; Bushnik, T. Examining the effects of a powered exoskeleton on quality of life and secondary impairments in people living with spinal cord injury. *Top. Spinal Cord Inj. Rehabil.* **2018**, *24*, 336–342. [CrossRef]
24. Jung, J.; Jang, I.; Riener, R.; Park, H. Walking intent detection algorithm for paraplegic patients using a robotic exoskeleton walking assistant with crutches. *Int. J. Control. Autom. Syst.* **2012**, *10*, 954–962. [CrossRef]
25. Jung, J.Y.; Heo, W.; Yang, H.; Park, H. A neural network-based gait phase classification method using sensors equipped on lower limb exoskeleton robots. *Sensors* **2015**, *15*, 27738–27759. [CrossRef]
26. Wang, S.; Wang, L.; Meijneke, C.; Van Asseldonk, E.; Hoellinger, T.; Cheron, G.; Ivanenko, Y.; La Scaleia, V.; Sylos-Labini, F.; Molinari, M. Design and control of the MINDWALKER exoskeleton. *IEEE Trans. Neural Syst. Rehabil. Eng.* **2015**, *23*, 277–286. [CrossRef]
27. Sanz-Merodio, D.; Puyuelo, G.; Ganguly, A.; Garces, E.; Garcia, E. EXOtrainer project clinical evaluation of gait training with exokeleton in children with spinal muscular atrophy. In *Advances in Robotics Research: From Lab to Market*; Springer: Berlin/Heidelberg, Germany, 2020.
28. Zoss, A.; Kazerooni, H.; Chu, A. On the Mechanical Design of the Berkeley Lowerextremity Exoskeleton (BLEEX). In Proceedings of the IEEE/RSJ International Conference on Intelligent Robots and Systems, Edmonton, AB, Canada, 2–6 August 2005; pp. 3465–3472.
29. Galle, S.; Malcolm, P.; Collins, S.H.; Clercq, D.D. Reducing the metabolic cost of walking with an ankle exoskeleton: Interaction between actuation timing and power. *J. Neuroeng. Rehabil.* **2017**, *14*, 35. [CrossRef]
30. Veale, A.J.; Xie, S.Q. Towards compliant and wearable robotic orthoses: A review of current and emerging actuator technologies. *Med. Eng. Phys.* **2016**, *38*, 317–325. [CrossRef] [PubMed]
31. Contreras-Vidal, J.L.; Grossman, R.G. NeuroRex: A Clinical Neural Interface Roadmap for EEG-Based Brain Machine Interfaces to a Lower Body Robotic Exoskeleton. In Proceedings of the 2013 35th Annual International Conference of the IEEE Engineering in Medicine and Biology Society, Osaka, Japan, 3–7 July 2013; pp. 1579–1582.
32. Kadone, H.; Kubota, S.; Abe, T.; Noguchi, H.; Miura, K.; Koda, M.; Shimizu, Y.; Hada, Y.; Sankai, Y.; Suzuki, K.; et al. Mucular activity modulation during post-operative walking with Hybrid Assistive Limb (HAL) in a patient with thoracic myelopathy due to ossification of posterior longitudinal ligament: A case report. *Front. Neurol.* **2020**, *11*, 102. [CrossRef] [PubMed]
33. Raj, A.K.; Neuhaus, P.D.; Moucheboeuf, A.M.; Noorden, J.H.; Lecoutre, D.V. Mina: A sensorimotor robotic orthosis for mobility assistance. *J. Robot.* **2011**, *2*, 284352.1–284352.8. [CrossRef]
34. Pransky, J. The Pransky interview: Russ Angold; Co-founder and president of Ekso(TM) labs. *Ind. Robot.* **2014**, *41*, 329–334. [CrossRef]
35. Suzuki, N.; Soga, T.; Izumi, R.; Toyoshima, M.; Shibasaki, M.; Sato, I.; Kudo, Y.; Aoki, M.; Kato, M. Hybrid Assistive Limb (R) for sporadic inclusion body myositis: A case series. *J. Clin. Neurosci.* **2020**, *81*, 92–94. [CrossRef]
36. Nankaku, M.; Tanaka, H.; Ikeguchi, R.; Kikuchi, T.; Miyamoto, S.; Matsuda, S. Effects of walking distance over robot-assisted training on walking ability in chronic stroke patients. *J. Clin. Neurosci.* **2020**, *81*, 279–283. [CrossRef]
37. Kuroda, M.; Nakagawa, S.; Mutsuzaki, H.; Mataki, Y.; Yoshikawa, K.; Takahashi, K.; Nakayama, T.; Iwasaki, N. Robot-assisted gait training using a very small-sized Hybrid Assistive Limb (R) for pediatric cerebral palsy: A case report. *Brain Dev.* **2020**, *42*, 468–472. [CrossRef]
38. Chen, G.; Chan, C.K.; Guo, Z.; Yu, H. A review of lower extremity assistive robotic exoskeletons in rehabilitation therapy. *Crit. Rev. Biomed. Eng.* **2013**, *41*, 343–363. [CrossRef]
39. Saito, Y.; Kikuchi, K.; Negoto, H.; Oshima, T.; Haneyoshi, T. Development of Externally Powered Lower Limb Orthosis with Bilateral-Servo Actuator. In Proceedings of the IEEE 9th International Conference on Rehabilitation Robotics, Chicago, IL, USA, 28 June–1 July 2005; pp. 394–399.
40. Ha, K.H.; Quintero, H.A.; Farris, R.J.; Goldfarb, M. Enhancing Stance Phase Propulsion during Level Walking by Combining FES with a Powered Exoskeleton for Persons with Paraplegia. In Proceedings of the 2012 Annual International Conference of the IEEE Engineering in Medicine and Biology Society, San Diego, CA, USA, 28 August–1 September 2012; pp. 344–347.
41. Xu, J.; Li, Y.; Xu, L.; Peng, C.; Chen, S.; Liu, J.; Xu, C.; Cheng, G.; Xu, H.; Liu, Y.; et al. A multi-mode rehabilitation robot with magnetorheological actuators based on human motion intention estimation. *IEEE Trans. Neural Syst. Rehabil. Eng.* **2019**, *27*, 2216–2228. [CrossRef]
42. Copaci, D.; Blanco, D.; Moreno, L.E. Flexible shape-memory alloy-based actuator: Mechanical design optimization according to application. *Actuators* **2019**, *8*, 63. [CrossRef]

43. Yamamoto, K.; Ishii, M.; Hyodo, K.; Yoshimitsu, T.; Matsuo, T. Development ofpower assisting suit (miniaturization of supply system to realize wearable suit). *JSME Int. J. Ser. C* **2003**, *46*, 923–930. [CrossRef]
44. Caldwell, D.G.; Tsagarakis, N.G.; Kousidou, S.; Costa, N.; Sarakoglou, I. "Soft" exoskeletons for upper and lower body rehabilitation–Design, control and testing. *Int. J. Hum. Robot.* **2007**, *4*, 549–573. [CrossRef]
45. Carvalho, A.; Navin, K.P.; Desai, V. Design and characterization of a pneumatic muscle actuator with novel end-fittings for medical assistive applications. *Sens. Actuators A Phys.* **2021**, *331*, 112877. [CrossRef]
46. Otten, A.; Voort, C.; Stienen, A.; Aarts, R.; van Asseldonk, E.; Kooij, H.V.D. LIMPACT: A hydraulically powered self-aligning upper limb exoskeleton. *IEEE Trans. Mechatron.* **2015**, *20*, 2285–2298. [CrossRef]
47. Zhao, G.; Sharbafi, M.; Vlutters, M.; van Asseldonk, E.; Seyfarth, A. Template model inspired leg force feedback based control can assist human walking. In Proceedings of the 2017 International Conference on Rehabilitation Robotics, London, UK, 17–20 July 2017; pp. 473–478.
48. Strausser, K.A.; Swift, T.A.; Zoss, A.B.; Kazerooni, H.; Bennett, B.C. Mobile Exoskeleton for Spinal Cord Injury: Development and Testing. In Proceedings of the Asme Dynamic Systems and Control Conference and Bath/Asme Symposium on Fluid Power and Motion Control, Bath, UK, 14–16 September 2012; pp. 419–425.
49. Choi, J.; Na, B.; Jung, P.-G.; Rha, D.-W.; Kong, K. WalkON suit: A medalist in the powered exoskeleton race of cybathlon. *2016 IEEE Robot. Autom. Mag.* **2017**, *24*, 75–86. [CrossRef]
50. Lenzi, T.; De Rossi, S.M.M.; Vitiello, N.; Carrozza, M.C. Intention-based EMG control for powered exoskeletons. *IEEE Trans. Biomed. Eng.* **2012**, *59*, 2180–2190. [CrossRef] [PubMed]
51. Athanasiou, A.; Lithari, C.; Kalogianni, K.; Klados, M.A.; Bamidis, P.D. Source detection and functional connectivity of the sensorimotor cortex during actual and imaginary limb movement: A preliminary study on the implementation of eCoonectome in motor imagery protocols. *Adv. Hum. Comput. Interact.* **2012**, *2012*. Available online: https://www.researchgate.net/publication/258381811 (accessed on 12 October 2020). [CrossRef]
52. Cheron, G.; Duvinage, M.; De Saedeleer, C.; Castermans, T.; Bengoetxea, A.; Petieau, M.; Seetharaman, K.; Hoellinger, T.; Dan, B.; Dutoit, T.; et al. From spinal central pattern generators to cortical network: Integrated BCI for walking rehabilitation. *Neural Plast.* **2012**, *2012*, 375148. [CrossRef] [PubMed]
53. Peruzzi, A.; Della Croce, U.; Cereatti, A. Estimation of stride length in level walking using an inertial measurement unit attached to the foot: A validation of the zero velocity assumption during stance. *J. Biomech.* **2011**, *44*, 1991–1994. [CrossRef] [PubMed]
54. Wang, L.; Sun, Y.; Li, Q.; Liu, T. Estimation of step length and gait asymmetry using wearable inertial sensors. *IEEE Sens. J.* **2018**, *18*, 3844–3851. [CrossRef]
55. Wang, L.; Sun, Y.; Li, Q.; Liu, T. Two shank-mounted imus-based gait analysis and classification for neurological disease patients. *IEEE Robot. Autom. Lett.* **2020**, *5*, 1970–1976. [CrossRef]
56. Li, W.Z.; Cao, G.Z.; Zhu, A.B. Review on Control Strategies for Lower Limb Rehabilitation Exoskeletons. *IEEE Access* **2021**, *9*, 123040–123060. [CrossRef]
57. Wang, H.; Mu, T.; Li, H.; Zhang, X. Research on Surface EMG Based Accurate Perception Method for Exoskeleton Robot Control. In Proceedings of the 2015 IEEE International Conference on Cyber Technology in Automation, Control and Intelligent Systems (IEEE-CYBER 2015), Shenyang, China, 9–12 June 2015; pp. 1900–1905.
58. Yatsuya, K.; Hirano, S.; Saitoh, E.; Tanabe, S.; Tanaka, H.; Eguchi, M.; Katoh, M.; Shimizu, Y.; Uno, A.; Kagaya, H. Comparison of energy efficiency between Wearable Power-Assist Locomotor (WPAL) and two types of knee-ankle-foot orthoses with a medial single hip joint (MSH-KAFO). *J. Spinal Cord Med.* **2018**, *41*, 48–54. [CrossRef] [PubMed]
59. Tanabe, S.; Koyama, S.; Saitoh, E.; Hirano, S.; Yatsuya, K.; Tsunoda, T.; Katoh, M.; Gotoh, T.; Furumoto, A. Clinical feasibility of gait training with a robotic exoskeleton (WPAL) in an individual with both incomplete cervical and complete thoracic spinal cord injury: A case study. *Neurorehabilitation* **2017**, *41*, 85–95. [CrossRef] [PubMed]
60. Griffin, R.; Cobb, T.; Craig, T.; Daniel, M.; van Dijk, N.; Gines, J.; Kramer, K.; Shah, S.; Siebinga, O.; Smith, J.; et al. Stepping forward with exoskeletons team IHMC's design and approach in the 2016 Cybathlon. *IEEE Robot. Autom. Mag.* **2017**, *24*, 66–74. [CrossRef]
61. Mummolo, C.; Peng, W.; Agarwal, S.; Griffin, R.; Neuhaus, P.; Kim, J. Stability of Mina v2 for robot-assisted balance and locomotion. *Front. Neurorobotics* **2018**, *12*, 16. [CrossRef]
62. Sanz-Merodio, D.; Sancho, J.; Perez, M.; Garcia, E. Control architecture of the ATLAS 2020 lower-limb active orthosis. *Adv. Coop. Robot.* **2017**, 860–868.
63. Riener, R.; Lünenburger, L.; Jezernik, S.; Anderschitz, M.; Colombo, G.; Dietz, V. Patient-cooperative strategies for robot-aided treadmill training: First experimental results. *IEEE Trans. Neural. Syst. Rehabil. Eng.* **2005**, *13*, 380–394. [CrossRef]
64. Veneman, J.F.; Kruidhof, R.; Hekman, E.E.G.; Ekkelenkamp, R.; Van Asseldonk, E.H.F.; van der Kooij, H. Design and evaluation of the LOPES exoskeleton robot for interactive gait rehabilitation. *IEEE Trans. Neural. Syst. Rehabil. Eng.* **2007**, *15*, 379–386. [CrossRef]
65. Brahmi, B.; Driscoll, M.; El Bojairami, I.K.; Saad, M.; Brahmi, A. Novel adaptive impedance control for exoskeleton robot for rehabilitation using a nonlinear time-delay disturbance observer. *ISA Trans.* **2021**, *108*, 381–392. [CrossRef]
66. Meuleman, J.; van Asseldonk, E.; van Oort, G.; Rietman, H.; van der Kooij, H. LOPES II—Design and evaluation of an admittance controlled gait training robot with shadow-leg approach. *IEEE Trans. Neural Syst. Rehabil. Eng.* **2016**, *24*, 352–363. [CrossRef] [PubMed]

67. Gu, Y.; Yang, D.; Huang, Q.; Yang, W.; Liu, H. Robust EMG pattern recognition in the presence of confounding factors: Features, classifiers and adaptive learning. *Expert Syst. Appl.* **2018**, *96*, 208–217. [CrossRef]
68. Lazarou, I.; Nikolopoulos, S.; Petrantonakis, P.C.; Kompatsiaris, I.; Tsolaki, M. EEG-based brain–computer interfaces for communication and rehabilitation of people with motor impairment: A novel approach of the 21st century. *Front. Hum. Neurosci.* **2018**, *12*, 14. [CrossRef] [PubMed]
69. Karavas, N.; Ajoudani, A.; Tsagarakis, N.; Saglia, J.; Bicchi, A.; Robotics, J.; Systems, A. Tele-impedance based assistive control for a compliant knee exoskeleton. *Robot. Auton. Syst.* **2015**, *73*, 78–90. [CrossRef]
70. Koller, J.R.; Jacobs, D.A.; Ferris, D.P.; Remy, C.D. Learning to walk with an adaptive gain proportional myoelectric controller for a robotic ankle exoskeleton. *J. Neuroeng. Rehabil.* **2015**, *12*, 97. [CrossRef]
71. Benabid, A.L.; Costecalde, T.; Eliseyev, A.; Charvet, G. An exoskeleton controlled by an epidural wireless brain–machine interface in a tetraplegic patient: A proof of concept demonstration. *Lancet Neurol.* **2019**, *18*, 1112–1122. [CrossRef]
72. Vouga, T.; Zhuang, K.Z.; Olivier, J.; Lebedev, M.A.; Nicolelis, M.A.L.; Bouri, M.; Bleuler, H. EXiO—A brain-controlled lower limb exoskeleton for rhesus macaques. *IEEE Trans. Neural Syst. Rehabil. Eng.* **2017**, *25*, 131–141. [CrossRef]
73. Gwin, J.T.; Ferris, D.P. An EEG-based study of discrere isometric and isotonic human lower limb muscle contractions. *J. Neuroeng. Rehabil.* **2012**, *9*, 35. [CrossRef]
74. Hong, E.; Gorman, P.H.; Forrest, G.F.; Asselin, P.K.; Knezevic, S.; Scott, W.; Wojciehowski, S.B.; Kornfeld, S.; Spungen, A.M. Mobility skills with exoskeletal-assisted walking in persons with SCI: Results from three center randomized clinical trial. *Front. Robot. AI* **2020**, *7*, 93. [CrossRef]
75. Norris, J.A.; Granata, K.P.; Mitros, M.R.; Byrne, E.M.; Marsh, A.P. Effect of augmented plantarflexion power on preferred walking speed and economy in young and older adults. *Gait Posture* **2007**, *25*, 620–627. [CrossRef]
76. Arazpour, M.; Bani, M.A.; Hutchins, S.W. Reciprocal gait orthoses and powered gait orthoses for walking by spinal cord injury patients. *Prosthet. Orthot. Int.* **2013**, *37*, 14–21. [CrossRef] [PubMed]
77. Lerner, Z.F.; Damiano, D.L.; Bulea, T.C.A. Robotic Exoskeleton to Treat Crouch Gait from Cerebral Palsy: Initial Kinematic and Neuromuscular Evaluation. In Proceedings of the 2016 38th Annual International Conference of the IEEE Engineering in Medicine and Biology Society (EMBC), Orlando, FL, USA, 16–20 August 2016; pp. 2214–2217.
78. Aguirre-Ollinger, G.; Colgate, J.E.; Peshkin, M.A.; Goswami, A. Inertia compensation control of a one-degree-of-freedom exoskeleton for lower-limb assistance: Initial experiments. *IEEE Trans. Neural Syst. Rehabil. Eng.* **2012**, *20*, 68–77. [CrossRef] [PubMed]
79. Li, W.; Lemaire, E.D.; Baddour, N. Design and Evaluation of a Modularized Ankle-Foot Orthosis with Quick Release Mechanism. In Proceedings of the 2020 42nd Annual International Conference of the IEEE Engineering in Medicine & Biology Society (EMBC), Montreal, QC, Canada, 20–24 July 2020; pp. 4831–4834.
80. Shao, Y.; Zhang, W.; Xu, K.; Ding, X. Design of a novel compact adaptive ankle exoskeleton for walking assistance. *Adv. Mech. Mach. Sci.* **2019**, *173*, 2159–2168.
81. Tan, C.; Sun, F.; Fang, B.; Kong, T.; Zhang, W. Autoencoder-based transfer learning in brain–computer interface for rehabilitation robot. *Int. J. Adv. Robot Syst.* **2019**, *16*, 1–12. [CrossRef]
82. Elia, N.D.; Vanetti, F.; Cempini, M.; Pasquini, G.; Parri, A.; Rabuffetti, M.; Ferrarin, M.; Molino Lova, R.; Vitiello, N. Physical human-robot interaction of an active pelvis orthosis: Toward ergonomic assessment of wearable robots. *J. NeuroEngineering Rehabil.* **2017**, *14*, 29.
83. Haldane, D.W.; Casarez, C.S.; Karras, J.T.; Lee, J.; Li, C.; Pullin, A.O.; Schaler, E.W.; Yun, D.; Ota, H.; Javey, A.; et al. Fearing Integrated manufacture of exoskeletons and sensing structures for folded millirobots. *J. Mech. Robot.* **2015**, *7*, 021011. [CrossRef]
84. Yandell, M.B.; Quinlivan, B.T.; Popov, D.; Walsh, C.; Zelik, K.E. Physical interface dynamics alter how robotic exosuits augment human movement: Implications for optimizing wearable assistive devices. *J. Neuroeng. Rehabil.* **2017**, *14*, 40. [CrossRef]
85. Martínez, A.; Lawson, B.; Goldfarb, M. A controller for guiding leg movement during overground walking with a lower limb exoskeleton. *IEEE Trans. Robot.* **2018**, *34*, 183–193.
86. Nagarajan, U.; Aguirre-Ollinger, G.; Goswami, A. Integral admittance shaping: A unified framework for active exoskeleton control. *Robot. Auton. Syst.* **2016**, *75*, 310–324. [CrossRef]
87. Aguirre-Ollinger, G.; Yu, H. Omnidirectional platforms for gait training: Admittance-shaping control for enhanced mobility. *J. Intell. Robot. Syst.* **2021**, *101*, 52. [CrossRef]
88. Zhang, T.; Tran, M.; Huang, H. Admittance shaping-based assistive control of SEA-driven robotic hip exoskeleton. *IEEE/ASME Trans. Mechatron.* **2019**, *24*, 1508–1519. [CrossRef]
89. Leal-Junior, A.G.; Frizera-Neto, A.; Pontes, M.J.; Botelho, T.R. Hysteresis compensation technique applied to polymer optical fiber curvature sensor for lower limb exoskeletons. *Meas. Sci. Technol.* **2017**, *28*, 125103. [CrossRef]
90. Gui, K.; Liu, H.; Zhang, D. A practical and adaptive method to achieve EMG-based torque estimation for a robotic exoskeleton. *IEEE/ASME Trans. Mechatron.* **2019**, *24*, 483–494. [CrossRef]
91. Wang, F.; Yan, L.; Xiao, J. Recognition of the gait phase based on new deep learning algorithm using multisensor information fusion. *Sens. Mater.* **2019**, *31*, 3041–3054. [CrossRef]
92. Cardona, M.; Cena, C.E.G.; Serrano, F.; Saltaren, R. ALICE: Conceptual development of a lower limb exoskeleton robot driven by an on-board musculoskeletal simulator. *Sensors* **2020**, *20*, 789. [CrossRef]

93. Krausz, N.E.; Hargrove, L.J. A Survey of Teleceptive Sensing for Wearable Assistive Robotic Devices. *Sensors* **2019**, *19*, 5238. [CrossRef]
94. Hussain, F.; Goecke, R.; Mohammadian, M. Exoskeleton robots for lower limb assistance: A review of materials, actuation, and manufacturing methods. *Proc. Inst. Mech. Eng. Part H J. Eng. Med.* **2021**, *235*, 1375–1385. [CrossRef]
95. Ronsse, R.; Lenzi, T.; Vitiello, N.; Koopman, B.; Van Asseldonk, E.; De Rossi, S.M.M.; Van Der Kieboom, J.; Van Der Kooij, H.; Carrozza, M.C.; Ijspeert, A.J. Oscillator-based assistance of cyclical movements: Model-based and model-free approaches. *Med. Biol. Eng. Comput.* **2011**, *49*, 1173–1185. [CrossRef]
96. Yu, H.; Huang, S.; Chen, G.; Pan, Y.; Guo, Z. Human–robot interaction control of rehabilitation robots with series elastic actuators. *IEEE Trans. Robot.* **2015**, *31*, 1089–1100. [CrossRef]
97. Liang, C.; Hsiao, T. Admittance control of powered exoskeletons based on joint torque estimation. *IEEE Access* **2020**, *8*, 94404–94414. [CrossRef]
98. Lv, G.; Zhu, H.; Gregg, R.D. On the design and control of highly backdrivable lower-limb exoskeletons: A discussion of past and ongoing work. *IEEE Control. Syst.* **2018**, *38*, 88–113. [CrossRef] [PubMed]
99. Rose, L.; Bazzocchi, M.C.F.; Nejat, G. End-to-End Deep Reinforcement Learning for Exoskeleton Control. In Proceedings of the IEEE International Conference on Systems, Man, and Cybernetics, Toronto, ON, Canada, 11–14 October 2020; pp. 4294–4301.

electronics MDPI

Article

Investigation of Input Modalities Based on a Spatial Region Array for Hand-Gesture Interfaces

Huanwei Wu [1], Yi Han [2], Yanyin Zhou [1], Xiangliang Zhang [3], Jibin Yin [1,*] and Shuoyu Wang [2]

1 Faculty of Information Engineering and Automation, Kunming University of Science and Technology, Kunming 650500, China; wuhuanwei@stu.kust.edu.cn (H.W.); zhouyy@stu.kust.edu.cn (Y.Z.)
2 School of System Engineering, Kochi University of Technology, Kochi 780-8515, Japan; 258012g@gs.kochi-tech.ac.jp (Y.H.); Wang.shuoyu@kochi-tech.ac.jp (S.W.)
3 State Key Laboratory of Fluid Power and Mechatronic Systems, School of Mechanical Engineering, Zhejiang University, Hangzhou 310027, China; xlzh@zju.edu.cn
* Correspondence: yjblovelh@aliyun.com; Tel.: +86-135-7707-3950

Abstract: To improve the efficiency of computer input, extensive research has been conducted on hand movement in a spatial region. Most of it has focused on the technologies but not the users' spatial controllability. To assess this, we analyze a users' common operational area through partitioning, including a layered array of one dimension and a spatial region array of two dimensions. In addition, to determine the difference in spatial controllability between a sighted person and a visually impaired person, we designed two experiments: target selection under a visual and under a non-visual scenario. Furthermore, we explored two factors: the size and the position of the target. Results showed the following: the 5 × 5 target blocks, which were 60.8 mm × 48 mm, could be easily controlled by both the sighted and the visually impaired person; the sighted person could easily select the bottom-right area; however, for the visually impaired person, the easiest selected area was the upper right. Based on the results of the users' spatial controllability, we propose two interaction techniques (non-visual selection and a spatial gesture recognition technique for surgery) and four spatial partitioning strategies for human-computer interaction designers, which can improve the users spatial controllability.

Keywords: target selection; spatial controllability; gesture recognition; spatial regions; visual and non-visual; regional division

Citation: Wu, H.; Han, Y.; Zhou, Y.; Zhang, X.; Yin, J.; Wang, S. Investigation of Input Modalities Based on a Spatial Region Array for Hand-Gesture Interfaces. *Electronics* **2021**, *10*, 3078. https://doi.org/10.3390/electronics10243078

Academic Editor: Paolo Visconti

Received: 28 October 2021
Accepted: 7 December 2021
Published: 10 December 2021

1. Introduction

In the field of human-computer interaction, there has been a wealth of related research on improving computer input efficiency, including voice input and gesture input, and among which interaction technologies based on common spaces and interaction technologies based on gesture recognition have also been commercialized and applied. Therefore, it is necessary to study the space operation range commonly used by users and combine the advantages of air gesture operation to obtain a new type of human-computer interaction input channel to improve the interaction experience and efficiency.

Spatial gestures and related recognition techniques have been widely used in various scenarios, such as intangible user interfaces and large-screen interactions. Generally, those gestures are to be executed by users in spatial areas within easy reach. Movement of the hands as a change of 3D positional data is an increasingly important input modality for computer interaction. In this mode, users often move their hands up and down in space to achieve a corresponding purpose.

1.1. Gesture Recognition

The field of gesture recognition has been a hot topic, with various potential applications from playing games to medical treatment. Different researchers have utilized various devices to conduct studies in this area. In terms of gesture data collection, common methods include: data gloves [1]; Kinect video capture devices [2]; Leap Motion capture devices [3]; collecting data through the device's first-view camera in the AR/VR environment [4,5]; and the use of heterogeneous sensors to collect data to improve the recognition rate [6]. In terms of experiment types, image segmentation [4] and image classification [2] are more common, and researching non-contact tactile feedback in AR/VR environments [5]. In [7], a cross-label recognition system is proposed. Promoting gesture recognition by improving large data intelligent editing processes [6]. There is also an identification method that measures the distance and angle between the fingers [1,8] studied arm gestures. Gesture recognition has many applications, such as gestures that interact with animation in shadow puppet shows [6] and interact with television [9]. In the area of medicine, doctors can use gestures to safely interact with computers to control images without the need to touch an operating room screen [10]. Navigating and manipulating large amounts of data suitable for high-resolution wall displays [11]. In the driving field, by exploring the space in front of an in-car screen, in-car touchscreen interaction can be expanded with the careful application of a target expansion strategy, allowing interaction with in-car systems to be more convenient [12]. Gesture recognition uses several technologies. In [1], the recognition rate is improved through a deep learning-based gesture spotting algorithm. In [4], a gesture recognition deep neural network was proposed which recognizes ego hand gestures from videos (videos containing a single gesture) by generating and recognizing embeddings of ego hands from image sequences of varying lengths. A novel deep neural network is designed in [7], which embeds gestures in the high-dimensional Euclidean space. It tackles the spatial resolution limits imposed by RF hardware and the specular reflection effect of RF signals. In [2], the support vector machine (SVM) classifier is used to classify the data. Ultrasonic haptic technology is used in E to develop and integrate air haptics that do not require wearing or holding any equipment in the virtual reality game experience [5]. This paper presents a set of spatial partitioning strategies for designers as guidelines that can improve the types of technologies described.

1.2. Interaction Based on a Spatial Region

Gesture interaction has developed rapidly as one of the important research areas of human-computer interaction. However, we have checked the existing literature and found that researchers are more concerned with new interactive technologies developed by interactive channels such as large screens, cameras, and sensors. These studies have made great contributions to improving the efficiency of human-machine interaction. Human activities and space are closely linked, so researchers must pay attention to the controllability capabilities of users' space. Interaction techniques based on a spatial region array are novel and promising and have a wide range of applications. To achieve multi-layer interaction, a novel multi-layered gesture recognition method using Kinect has been proposed and explores the essential linguistic characters of gestures [13]. The method can obtain relatively high performance. Multi-layer interaction techniques divide the interaction space into multiple interaction layers. Each layer has a special function; users can access different commands by accessing the different layers. The overall interaction height and different minimum layer thicknesses for vertical and horizontal search tasks were experimentally explored in [14]. In [15], three target selection techniques were developed for air pointing: small angular ray casting movements, large movements in a 2D plane, and movements in a 3D volume. Although those techniques were designed systematically to use from one to three dimensions, the target selection techniques were presented without strategies of common space partitioning. Many researchers have designed techniques based on spatial regions, but they have not focused on the division of space [16,17]. Some researchers have tried to divide the space using angles [18–22]. However, there is a lack of basic research on common spatial partitioning. The purpose of this paper is to explore common operational spatial partitioning in the user interface.

1.3. Interaction of Visually Impaired Individuals

There is a need for computer interactions that can also be used by visually impaired individuals; meeting this need has appealed to many researchers. A framework was proposed for exploring the differences between the spatial sense ability of visually impaired and sighted persons in three longitudinal models [23]. Through exploring the effect of spatial ability on a visually impaired person's sense of position within web pages, we know users can obtain an accurate overview of a web page with audio feedback when using a touchscreen [24]. By connecting the use of touch sensation and other multimedia design elements, it was found that touch sensation plays a critical role in improving application design for people with visual impairments [25]. Although there is a lack of systematic study on the common operational spatial region array of a visually impaired individual, gesture-free interaction by the status of thumb (GIST) is a wearable gestural interface that uses a depth camera to collect a users' hand gestures and can help a visually impaired individual perform everyday tasks [26]. There are techniques based on the two-dimensional structures of a keyboard surface that explore different methods of non-visual interaction [27]. To enable the blind to read the text, an affordable mobile application for the visually impaired person was proposed. The text could be read into speech format using text-to-speech conversion in a Text to Speech (TTS) framework [28]. Immersive virtual reality (VR) to provide a realistic walking experience for the visually impaired is proposed in [29]. A novel immersive interaction using a walking aid, i.e., a white cane, is designed to enable users with visual impairments to process ground recognition and inference processes realistically.

In summary, interactive technology based on spatial gestures has been integrated into people's daily lives, including visual users and visually impaired people. Therefore, further research on interaction technology based on spatial gestures is beneficial to improve the interaction efficiency between users and computers in daily life.

2. Materials and Methods

The extensive research mentioned above has focused on design techniques. However, this paper focuses on developing a set of guidelines based on spatial partitioning strategies. To assess users' spatial controllability, we attempt to reveal the common operational region when executing spatial gestures. Thus, in this paper, we have focused on investigating input modalities based on a spatial region array for hand gesture interfaces. We conducted a systematic study of human performance when selecting targets with a spatial region array, and developed two interaction techniques and four spatial partitioning strategies as design guidelines for human-computer interaction designers.

A Leap Motion M010 controller, a computer (including a keyboard and a display screen), and an experimental model designed by Unity 3D in the C# language were used in the experiment. The Leap Motion device can detect the hand's position in a range from 25° to 165° and is symmetrical. The experimental program was designed in Visual Studio 2019 and the Unity 3D Environment and ran on a 3.60 GHz AMD Ryzen R5-3600 CPU PC with Windows 10 Professional. The display resolution was set to 1000 × 800 pixels in the pilot studies and 1920 × 1080 pixels in Experiment 1 and Experiment 2.

To improve the users' spatial controllability, we first focused on the height and width of a rectangle (in front of and parallel to the screen) representing the average range of hand movements when a user sits down at a desk. We first determined the common operation area through a pilot study, which was realized by Leap Motion and unity 3D, as shown in Figure 1a. Leap Motion systems can detect and track hands, fingers, and finger-like tools. Its visual range is an inverted pyramid with the spire in the center of the equipment, as shown in Figure 1b. Leap motion's system adopts the right-hand Cartesian coordinate system, and the returned values are in real-world millimeters. The origin is at the center of the leap motion controller. The x-axis and z-axis are on the horizontal plane of the device, the x-axis is parallel to the long side of the device, the z-axis is parallel to the short side, and the y-axis is vertical upward, as shown in Figure 1c. Leap motion provides a set of dataset updates, and each frame of data contains a list of basic tracking data. When a hand is detected, it is assigned a unique ID indicator. For as long as the motion is analyzed, the leap motion program will give the frame motion factors based on the motion of the hand. Through the hand object, the current position information of the hand can be obtained. Unity 3D is a tool for creating interactive applications. It adopts a graphical development environment and can deploy projects to multiple platforms such as Windows. Unity's coordinates are world coordinates, which are consistent with leap motion, so we can accurately locate the hand motion in unity's world coordinates.

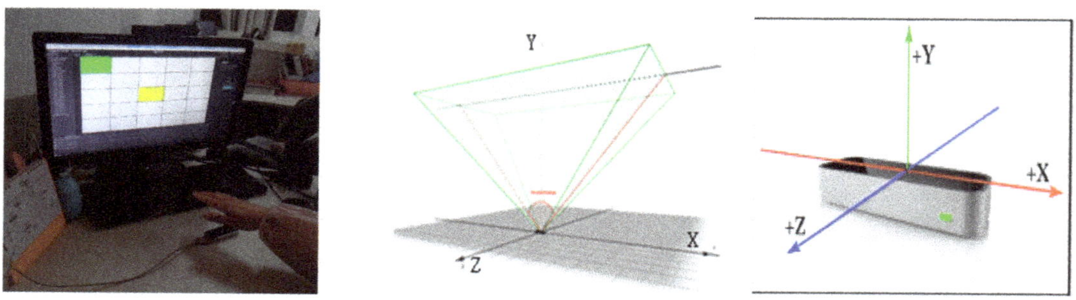

(a) (b) (c)

Figure 1. Schematic figure of experimental process and equipment: (**a**) experimental process, including pilot study, experiment 1 and 2; (**b**) the detectable spatial area of Leap Motion; (**c**) coordinate system of Leap Motion.

We imported a toolkit that supports Leap Motion gesture development in Unity. The toolkit contains prefabricated hands, related gesture action scripts, and case demonstrations, all of which can be used to help developers complete Leap Motion development work. The next step was to build an experimental development platform and add the "LeapHandController" prefab to the created scene. By observing whether the hand on the interface was within the capture range of the camera, we adjusted it to a suitable position and adjusted the size of the hand controller. The parameter was set to 1 to make it the same size as the real hand, so that the real hand could be moved in real-time to control the movement of the virtual hand, which is convenient for the user interaction operation described later. We imported the "Vectrosity" plug-in to meet the interface drawing requirements in our experiment. In this experiment, the plug-in was used to edit the experimental interface and achieve dynamic performance (for example, green represented a random target; when the target was selected, it appeared red; and yellow indicated the movement trajectory of the hand, etc.). To achieve the purpose of collecting experimental data, we recorded the acquired data in an Excel file and saved the file to the local disk. The logic processing of the business was implemented by C#. The logic included the method of drawing rectangles, the method of drawing UI interface, the method of setting the timer, the method of deleting rectangles, the method of randomly generating non-repeated layers, the method of setting data table, and the method of writing data to the Excel table.

2.1. Pilot Study

The study focused on designing, conducting, and analyzing a users' performance on a spatial region array, and addressed the following issues:

- Finding the physical limits of the common operational spatial region in the vertical and horizontal direction and setting this region as a study object.
- Finding the threshold of the target size levels when the users accomplish the target selection tasks under the visual and non-visual scenarios.
- Finding the relationship between the target regions when the users accomplish the target select operations under the visual and non-visual scenarios.

Many possible factors impact the interaction between the users and the Leap Motion controller. For example, the size of the spatial region array, the sensitivity of the Leap Motion device, a visual or non-visual task, and whether the users' performance of the task used the left or right hand. For the study's manageability and validity, we restricted our investigation to a situation where users sat in front of the Leap Motion device, centering it between the computer screen and the users' body, as Figure 1a shows.

2.1.1. Participants

Twelve students (two females, 10 males) participated in the user study. Their ages ranged from 22 to 30 (M = 25, SD = 2.08). The average body height was 168.17 cm (SD = 8.96). All of them were daily computer users.

2.1.2. Task & Procedure

To test the users' horizontal and vertical common range, we set visual cues to let the user move their hands horizontally and vertically while selecting the target block. First, the user pressed the "Start" button, and a green target block appeared randomly in a horizontal or vertical direction, as shown in Figure 2a,b. Taking the spatial position corresponding to the target block as a reference, when the user's hand moved to the spatial position in front of the screen corresponding to the target block, the block would turn red. A second target block would then randomly appear. In this and subsequent target selection tasks, the target block was always displayed in red when a gesture was made toward it. When the user thought their hand overlapped the target block, they could select the target by pressing the left Ctrl button. At the same time, the position of the users' hand was saved into an Excel file, and the next target selection task began.

(**a**) (**b**)

Figure 2. The beginning interface of the pilot experiment: (**a**) the horizontal experiment; (**b**) the vertical experiment.

Prior to the formal experiment, the participants were allowed to warm up by practicing until they understood and performed the task correctly. Taken together, these two experiments included the following: 12 subjects × 8 block levels × 14 repetitions × 2 directions = 2688 target selection trials.

2.1.3. Result

We analyzed the frequency of the user's hand position at each interval. We then were able to reach a conclusion regarding the user's vertical and horizontal common range, which was used as the study object of the following experiments.

After analyzing the collected data, we found that individuals manipulated their hands horizontally within the range (−220, 240) and vertically over the range (30, 370). The effective horizontal range was an interval of (−190, +190), and the vertical range was an interval of (50, 350), as Figure 3a,b show. We then chose 80% of this interval as the most common operational range, which was defined as a rectangle. The vertical interval was (80, 320), and the horizontal interval was (−152, +152). So the common operational region was an area of 240 mm × 304 mm, located 80 mm above the desktop.

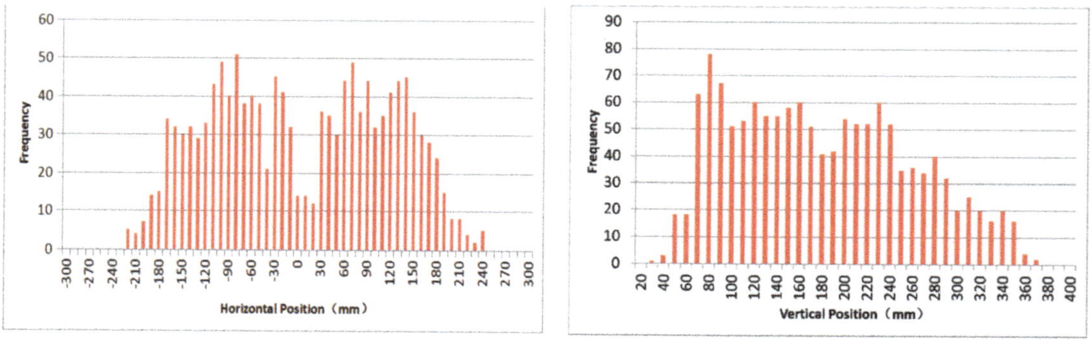

(a) (b)

Figure 3. The results of the pilot study: (**a**) the results of the horizontal experiment; (**b**) the results of the vertical experiment.

Based on these results, we set this common operational region as a study object and divided it into differently sized target arrays and target regions through even spatial partitioning. The common operational region was divided into four evenly sized sections, as shown in Table 1. There were six target-size levels, as shown in Table 2. We then designed experiments in which the participants attempted target selection tasks at different target-size levels, and with target positions within the different regions.

Table 1. The position of each region.

Region	A	B	C	D
Position	Upper Left	Upper Right	Bottom Left	Bottom Right

Table 2. The target size (length × width) at each level of the experimental condition.

Level	3 × 3	4 × 4	5 × 5	6 × 6	7 × 7	8 × 8
Size (mm)	101.33 × 80	76 × 60	60.8 × 48	50.7 × 40	43.43 × 34.29	38 × 30

In addition, differences between sighted and visually impaired individuals were also considered. The participants finished target selection tasks under both visual and non-visual scenarios. We then analyzed the data collected, including the average time and error rate.

The contributions of this work are:

1. It improves understanding the users and their controllability of space by identifying the common spatial region of users and the thresholds of target size and position.
2. We proposed two interaction techniques and four interaction strategies concerning the target size and position in the spatial region.

3. Experiment 1: Visual Scenario

To test the users' performance accuracy when conducting the interaction task, we set a visual cue for sighted users.

3.1. Participants & Apparatus

The participants and apparatus in Experiment 1 were the same as in the pilot study.

3.2. Task & Procedure

In Experiment 1, the current position of the users' hand mapping to the current block was shown on the experiment screen in real time. The target block turned from green to red while the current block overlapped with it, as shown in Figure 4. The user sat in a chair at the desk before the computer as they did in the pilot study.

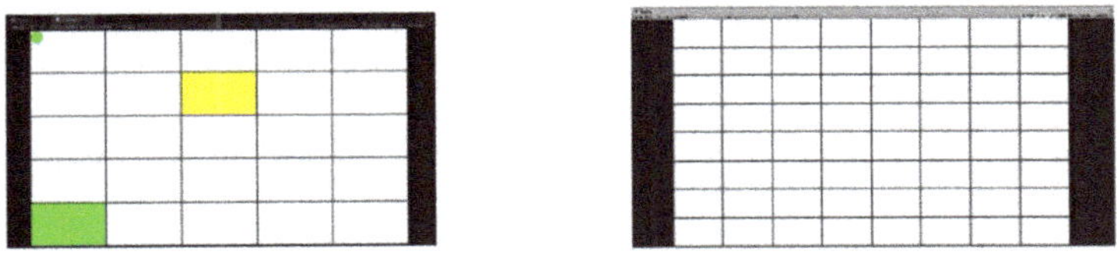

(a) (b)

Figure 4. The experiment interface of Experiments 1 and 2: (**a**) experiment interface of the 5 × 5 level; (**b**) experiment interface of the 8 × 8 level.

To maintain consistency in the experimental data, we saved the data to an Excel file at the end of a single target selection task, which was the moment the left Ctrl button was pressed. In addition, the experiment recorded a standard timestamp for an incremental time in the Unity 3D program. Once the user pressed the left Ctrl button to complete a target, the next target selection task began timing. To ensure an equal time for task selection, the user always placed their left index finger on the left Ctrl button. Before the formal experiment, participants were allowed a warm-up practice session until they could understand and perform the task correctly. In total, the experiment consisted of the following: 12 subjects × 6 target size levels × 4 target regions × 2 blocks × 3 repetitions = 1728 target selection trials.

3.3. Results

3.3.1. Selection Time

In the process of the experimental data analysis, we set the target level (3 × 3, 4 × 4, 5 × 5, 6 × 6, 7 × 7, and 8 × 8) and the target region (A, B, C, and D) as independent variables. In this way, we performed repeated measurements ANOVAs ($\alpha = 0.05$) on the time and accuracy of the target selection. The target selection time was defined as beginning from when the user clicked the Start button or pressed the left Ctrl button to when the user pressed the left Ctrl button again.

There was a main effect on the average time of the different regions ($F_{2.058, 22.634} = 11.460$, $p < 0.001$), see Figure 5a. The post hoc tests showed that there were no significant differences among the regions ($p > 0.05$) except for between regions B and C ($p = 0.035$) and regions B and D ($p < 0.001$). Region B had the fastest completion time, and region D had the slowest completion time.

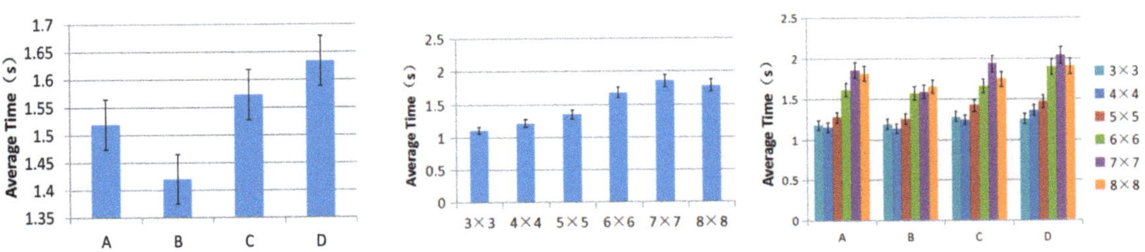

(a) (b) (c)

Figure 5. The average time of Experiment 1. The error bars represent a 95% confidence interval: (**a**) average selection time with different regions; (**b**) average selection time with different target size levels; and (**c**) average selection time for different target size levels and different regions (A, B, C, and D).

There was a main effect for the average time of the different levels of target size ($F_{2.321, 25.531} = 20.714$, $p < 0.001$), see Figure 5b. The post hoc tests showed that the shortest time was for the 3 × 3 level, and the longest was for the 7 × 7 level. There were no significant differences between the 3 × 3, 4 × 4, and 5 × 5 levels ($p > 0.119$). There were no significant differences between the 6 × 6, 7 × 7, and 8 × 8 levels ($p > 0.05$).

Further analysis of the level of target size × target region on selection time showed there was no significant interaction ($F_{4.272, 46.992} = 1.601$, $p = 0.187$), see Figure 5c. When the users selected a target in the 3 × 3 level, the shortest selection time was needed on average, while the 7 × 7 level had the longest time.

3.3.2. Selection Error Rate

The percentage of trials in which subjects made erroneous selections was defined as the selection error rate.

As shown in Figure 6a, there was a main effect on the average error rate of the different regions ($F_{3, 33} = 4.240$, $p = 0.012$). Post hoc tests showed no significant differences among all the regions ($p > 0.052$), except between regions B and D ($p = 0.033$). Region D had the lowest completion error rate, and region C had the highest completion error rate.

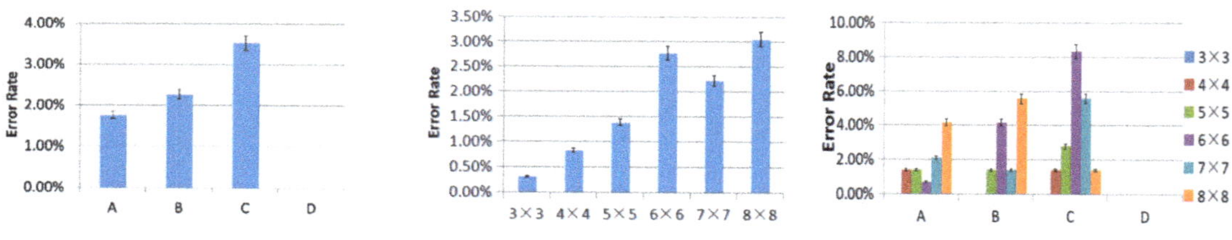

Figure 6. The error rate of Experiment 1. The error bars represent a 95% confidence interval: (**a**) average error rate with different regions; (**b**) average error rate with different target size levels; and (**c**) average error rate for different target size levels under different regions (A, B, C, and D).

As shown in Figure 6b, different target sizes had no significant effect on the average error rate ($F_{2.402,26.419} = 2.617$, $p = 0.083$). The post hoc tests showed no significant differences among all the target size levels ($p > 0.157$). The 3 × 3 level had the lowest error rate, and the 8 × 8 level had the highest. The higher the target size level, the higher the error rate when selecting the target. The largest increase in the error rate for adjacent levels was from the 5 × 5 to the 6 × 6 level. Thus, a target size of 5 × 5 (60.8 mm length and 48 mm width) provided a threshold for the most selections without a noticeable change in error rate.

Further analysis of what effect target size level × target region had on selection error rate showed there was no significant interaction ($F_{4.256, 46.816} = 1.665$, $p = 0.171$), see Figure 6c. The 3 × 3 target size level had the lowest selection error rate. The second-lowest selection error rate was the 4 × 4 level. The 8 × 8 level produced the highest error rate. The participants had the lowest error rate (0%) when the target region was D (bottom-right corner) and the highest error rate (3.54%) when the target region was C (bottom-left corner).

In previous literature [19], the author studied pointing at virtual buttons. The space is divided into 5 different sizes according to the angle, that is, the number of buttons. The experimental results show that the error rates are 0, 3.6%, 2.2%, 16.0%, 3.2%, respectively. As shown in Figure 6a, the error rates of our results are 0.31%, 0.83%, 1.39%, 2.78%, 2.22%, 3.06%, and the overall error rate is better. We also divided the regions, discussed the situation of each region, and the comprehensive situation of region and size. The literature only considers the error rate and not the task completion time. We comprehensively analyze the error rate and time and give suggestions for designing interactive technologies based on spatial regions, which are more convincing. Next, we studied the division of spatial regions in the absence of vision and give suggestions for designing interactive technologies based on spatial regions in the absence of vision.

3.4. Comparative Experiment 1

The participants in Experiment 1 were the same as in the pilot study, so they were trained and familiar with the experiment. To eliminate this influence, we invited 12 external participants who didn't know the experiment in advance. The experiment process was the same as experiment 1.

3.4.1. Selection Time

There was a main effect on the average time of the different regions ($F_{2.058, 22.634} = 11.168$, $p < 0.001$), see Figure 7a. The post hoc tests showed that there were no significant differences among the regions ($p > 0.05$) except for between regions B and C ($p = 0.023$) and regions B and D ($p = 0.002$). Region B had the fastest completion time, and region D had the slowest completion time.

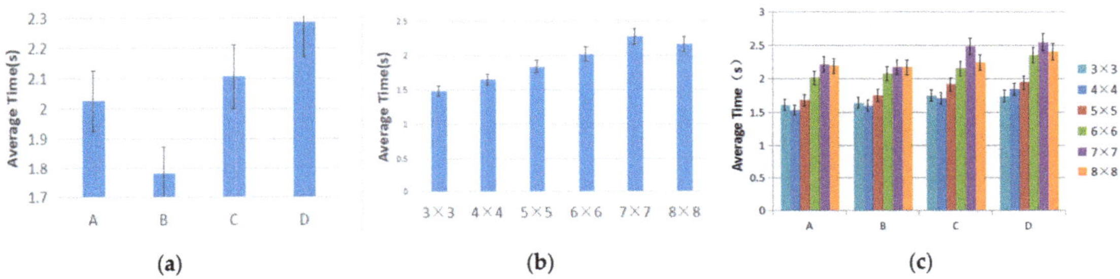

Figure 7. The average time of comparative Experiment 1. The error bars represent a 95% confidence interval: (**a**) average selection time with different regions; (**b**) average selection time with different target size levels; and (**c**) average selection time for different target size levels and different regions (A, B, C, and D).

There was a primary effect regarding the average time of the different levels of target size ($F_{2.321, 25.531} = 20.870$, $p < 0.001$), see Figure 7b. The post hoc tests showed that the shortest time was for the 3×3 level, and the longest was for the 7×7 level. There were no significant differences among the 3×3, 4×4, and 5×5 levels ($p > 0.37$). There were no significant differences among the 6×6, 7×7, and 8×8 levels ($p > 0.05$).

Further analysis of the level of target size × target region on selection time showed there was no significant interaction ($F_{4.272, 46.992} = 1.635$, $p = 0.178$), see Figure 7c. When the users selected a target in the 3×3 level, the shortest selection time was needed on average, while the 7×7 level had the longest time.

3.4.2. Selection Error Rate

As shown in Figure 8a, there was a primary effect on the average error rate of the different regions ($F_{3, 33} = 3.996$, $p = 0.016$). Post hoc tests showed no significant differences among all the regions ($p > 0.063$). Region D had the lowest completion error rate, and region C had the highest completion error rate.

As shown in Figure 8b, there was no significant effect for the average error rate of the different target sizes ($F_{2.109, 23.199} = 2.184$, $p = 0.133$). The post hoc tests showed no significant differences among all the target size levels ($p > 0.122$). The 3×3 level had the lowest error rate, and the 8×8 level had the highest.

Further analysis of target size level × target region on selection error rate showed there was no significant interaction ($F_{4.9, 53.903} = 1.628$, $p = 0.17$), see Figure 8c.

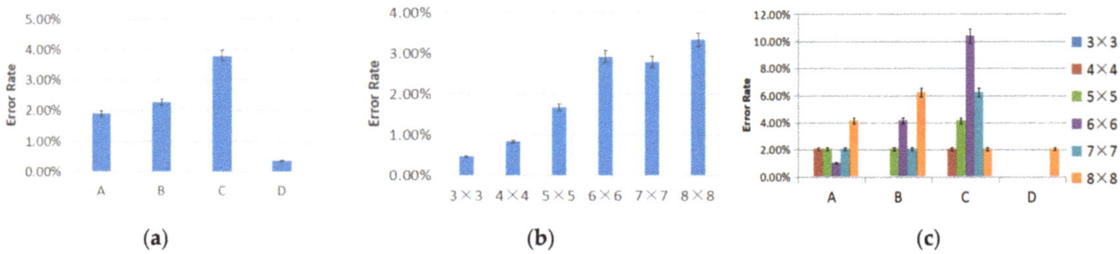

Figure 8. The error rate of comparative Experiment 1. The error bars represent a 95% confidence interval: (**a**) average error rate with different regions; (**b**) average error rate with different target size levels; and (**c**) average error rate for different target size levels under different regions (A, B, C, and D).

Compared with Experiment 1, the error rate of this experiment was slightly higher, and the average time was slightly longer, caused by the fact that new participants were not familiar with the experiment. The results showed that the regions (and levels) with the highest or lowest error rates were the same as Experiment 1. The regions (and levels) with the fastest or slowest average time were the same as Experiment 1, as shown in Figures 7 and 8.

4. Experiment 2: Non-Visual Scenario

To test the accuracy when the users performed the task with an eyes-free scenario, we set voice guidance for a visually impaired individual. The position of the users' hand mapping to the current block was shown on the experiment screen in real time, and the target block turned from green to red while the current block overlapped it.

4.1. Participants & Apparatus

The participants and apparatus in Experiment 2 were the same as in the pilot study.

4.2. Task & Procedure

The design and tasks were almost the same as in Experiment 1. The difference in Experiment 2 was that there was no visual feedback for the users, only voice guidance. The beginning guide audio was "The target block is X", and it would then announce the number of the block of the users' hand in real time. The participants already knew the number of the target block. This cycle would continue until the task was completed, and then the audio would announce, "This round of the experiment ends". Before the formal experiment, participants were allowed to warm up with a practice session until they could understand and perform the task correctly. In total, the experiment consisted of the following: 12 subjects × 6 target size levels × 4 target regions × 2 blocks × 3 repetitions = 1728 target selection trials.

4.3. Results

4.3.1. Selection Time

We found a main effect on the average time of different regions ($F_{3,33} = 13.496$, $p < 0.001$), see Figure 9a. The post hoc tests showed a significant difference between regions A and C ($p = 0.007$) and regions A and D ($p = 0.002$). There was a significant difference between regions B and C ($p = 0.042$) and regions B and D ($p = 0.002$). Other regions had no significant differences ($p > 0.975$). Region B had the fastest completion time, and region D had the slowest.

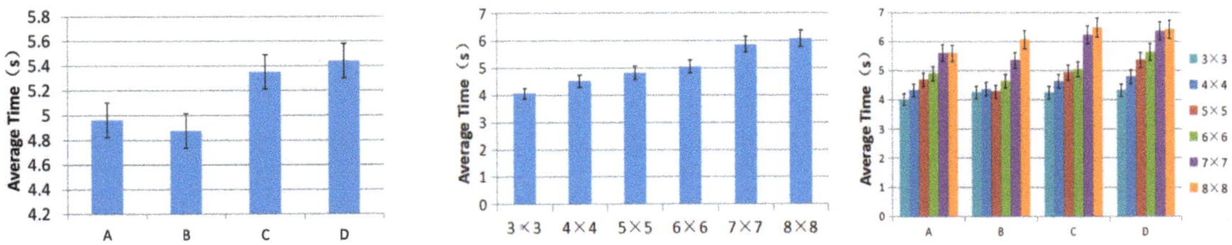

(a) (b) (c)

Figure 9. Average times of Experiment 2. Error bars represent a 95% confidence interval: (**a**) average selection time with different regions; (**b**) average selection time with different target size levels; and (**c**) average selection time for different target size levels under different regions (A, B, C, and D).

As shown in Figure 9b, there was a primary effect on the average time of the different target sizes ($F_{2.222, 24.445} = 24.893$, $p < 0.001$). A post hoc test showed no significant difference between the 3 × 3 and 4 × 4 levels ($p = 0.231$). There was no significant difference between the 5 × 5 and 6 × 6 levels ($p = 0.270$). There was no significant difference between the 7 × 7 and 8 × 8 levels ($p = 0.142$). The 3 × 3 level had the fastest completion time, and the 8 × 8 level had the slowest completion time. The higher the target size level, the longer the time needed to select the target. The largest increase in selection time for adjacent levels was from the 6 × 6 to the 7 × 7 level. Thus, the target size of 6 × 6 (50.67 mm length and 40 mm width) provided a threshold of the most selections without a noticeable change in selection time.

Further analysis of target size level × target region on selection time showed there was no significant interaction ($F_{4.912, 54.027} = 1.575$, $p = 0.184$), see Figure 9c. The shortest selection time was needed when users selected a target in the 3 × 3 level. The second shortest level was the 4 × 4 level, while the 8 × 8 level took the longest time.

4.3.2. Selection Error Rate

There was no significant effect concerning the average error rate of the different regions ($F_{3, 33} = 0.909$, $p = 0.447$), see Figure 10a. The post hoc tests showed no significant differences among all the regions ($p = 0.670$). Region B had the lowest completion error rate, and region C had the highest.

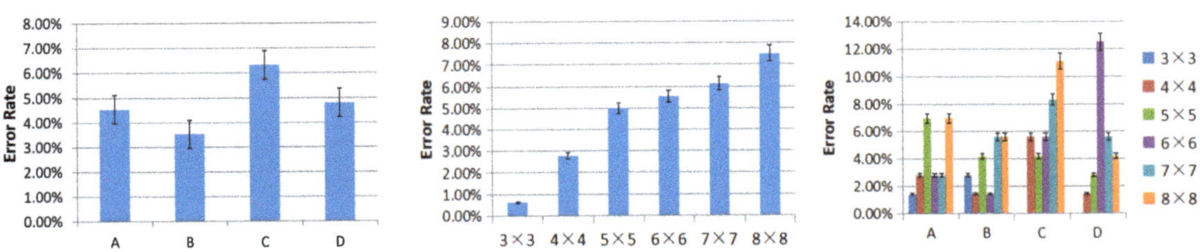

(a) (b) (c)

Figure 10. The error rate of Experiment 2. The error bars represent a 95% confidence interval: (**a**) average error rate with different regions; (**b**) average error rate with different target size levels; (**c**) average error rate for different target size levels under different regions (A, B, C, and D).

As shown in Figure 10b, there was no significant effect concerning the average error rate of the different target sizes ($F_{5, 55} = 4.388$, $p = 0.002$). The post hoc tests showed no significant differences among all the levels of target size ($p > 0.156$), except for the 3 × 3 level and 6 × 6 level ($p = 0.030$). The 3 × 3 level had the lowest completion error rate, and the 8 × 8 level had the highest completion error rate. The higher the target size level, the higher the error rate needed to select a target. The largest increase in the selection error rate for adjacent levels was from the 4 × 4 to the 5 × 5. Thus, a target size of 4 × 4 (76 mm length and 60 mm width) provided a threshold of the most selections without a noticeable change in selecting error rate.

Further analysis of the target size level × target region on selection error rate showed there were no significant interaction ($F_{5.122,\,56.337} = 1.524$, $p = 0.196$), see Figure 10c. On average, the 3 × 3 target size level had the lowest selection error rate. The second-lowest selection error rate was for the 4 × 4 level. The 8 × 8 level produced the highest error rate. The participants reached the lowest selection error rate (3.54%) when the target region was B (upper-right corner), and the highest error rate (6.31%) when the target region was C (bottom-left corner).

4.4. Comparative Experiment 2

The participants in Experiment 2 were the same as in the pilot study, so they were trained and familiar with the experiment. To eliminate this influence, we invited 12 external participants who did not know the experiment in advance. The experiment process was the same as experiment 2.

4.4.1. Selection Time

We found a main effect on the average time of different regions ($F_{3,\,33} = 13.474$, $p < 0.001$), see Figure 11a. The post hoc tests showed a significant difference between regions A and C ($p = 0.007$) and regions A and D ($p = 0.003$). There was a significant difference between regions B and C ($p = 0.042$) and regions B and D ($p = 0.002$). Other regions had no significant differences among them ($p > 0.994$). Region B had the fastest completion time, and region D had the slowest completion time.

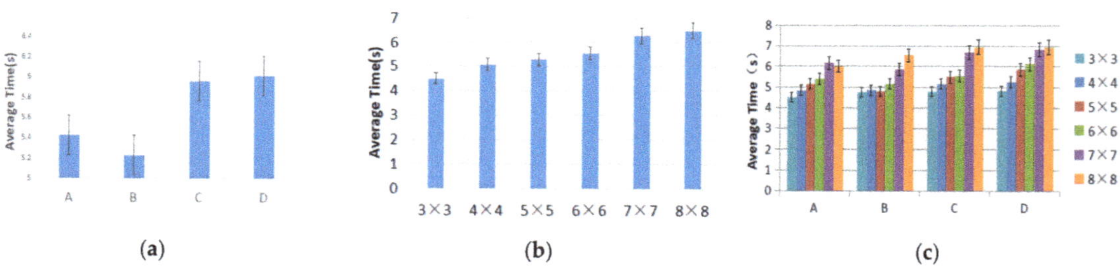

Figure 11. The average time of comparative Experiment 2. The error bars represent a 95% confidence interval: (**a**) average selection time with different regions; (**b**) average selection time with different target size levels; and (**c**) average selection time for different target size levels and different regions (A, B, C, and D).

As shown in Figure 11b, there was a main effect on the average time of the different target sizes ($F_{2.222,\,24.445} = 25.009$, $p < 0.001$). A post hoc test showed no significant difference between the 3 × 3 and 4 × 4 levels ($p = 0.227$). There was no significant difference between the 5 × 5 and 6 × 6 levels ($p > 0.05$). There was no significant difference between the 7 × 7 and 8 × 8 levels ($p > 0.05$). The 3 × 3 level had the fastest completion time, and the 8 × 8 level had the slowest completion time.

Further analysis of target size level × target region on selection time showed there was no significant interaction ($F_{4.912,\,54.027} = 1.574$, $p = 0.184$), see Figure 11c.

4.4.2. Selection Error Rate

There was no significant effect for the average error rate of the different regions ($F_{3,\,33} = 0.883$ $p = 0.46$), see Figure 12a. The post hoc tests showed no significant differences among all the regions ($p > 0.05$). Region B had the lowest completion error rate, and region C had the highest completion error rate.

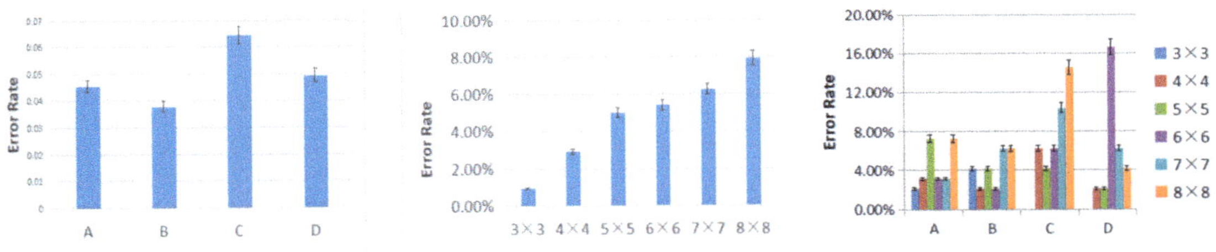

(a) (b) (c)

Figure 12. The error rate of comparative Experiment 1. The error bars represent a 95% confidence interval: (**a**) average error rate with different regions; (**b**) average error rate with different target size levels; and (**c**) average error rate for different target size levels under different regions (A, B, C, and D).

As shown in Figure 12b, there was no significant effect for the average error rate of the different target sizes ($F_{5,55} = 3.193$, $p = 0.013$). The post hoc tests showed no significant differences among all the levels of target size ($p > 0.109$). The 3 × 3 level had the lowest completion error rate, and the 8 × 8 level had the highest completion error rate.

Further analysis of the target size level × target region on selection error rate showed there were no significant interaction ($F_{6.461,71.069} = 1.085$, $p = 0.381$), see Figure 12c. Compared to Experiment 2, the error rate of this experiment was slightly higher, and the average time was slightly longer, caused by the fact that new participants were not familiar with the experiment. The results showed that the regions (and levels) with the highest or lowest error rates were the same as in Experiment 2. The regions (and levels) with the fastest or slowest average time were the same as Experiment 2, as shown in Figures 11 and 12.

5. Discussion & Conclusions

In this work, we analyzed the users' common operational area regarding partitioning and the difference in spatial controllability between a sighted and a visually impaired individual. We introduced three experiments and a pilot study concerning the common spatial range and the thresholds of the target size level in the spatial region for a sighted and visually impaired individual. We compared the speed and accuracy of the target dimensions of six different levels and the difference in speed and accuracy among the four azimuth regions of A, B, C, and D in both visual and non-visual scenarios. Many of our performance study results were statistically significant, which allows us to draw many meaningful conclusions about human-computer interaction in spatial regions that can be used for designing techniques for sighted and visually impaired individuals. This paper focused on systematically analyzing the common operational range of one dimension and the threshold of two dimensions. The results are as follows:

- Common operational range. As a result of the pilot study, the horizontal range of the common operational range was the interval of (−152, +152), and the vertical range was the interval of (80, 320), which means that the rectangle's length was 304 mm, and the width was 240 mm.
- Threshold of target size levels. For a sighted person, the threshold target size was the 5 × 5 level, whose length was 60.8 mm, and the width was 48 mm. For a visually impaired individual, the threshold target size was the 4 × 4 level, whose length was 76 mm, and the width was 60 mm.
- Target region thresholds. For a sighted individual, the threshold target regions were region B with the shortest selection time, and region D with the lowest selection error rate. There was a significant difference between regions B and D. For a visually impaired individual, the threshold target region with the shortest selection time was region B. There were no significant differences in the error rate of target selection among all the levels.

Based on the above results and findings, we have developed a set of preliminary guidelines regarding target selection in spatial partitioning scenarios:

1. For visual scenarios, Region D (the bottom-right corner) is not recommended when high speed in selecting a target is needed. However, due to its lower error rate, region D remains a good alternative for scenes with higher requirements for a correct rate.
2. For visual scenarios, the 6 × 6 level (50.7 mm × 40 mm) is not recommended because both its error rate and the average time were high. We recommend the 5 × 5 level (60.8 mm × 48 mm) after considering the selection time and error rate.
3. For non-visual scenarios, we recommend region B (the upper-right corner) after considering the selection time and error rate. Another reason is that region B costs the least time and error rate when selecting targets.
4. For non-visual scenarios, we recommend the 4 × 4 level (76 mm × 60 mm) after considering the selection time and error rate because it served as the threshold.

In the case of vision, researchers can refer to the design suggestions in Table 3 when studying the space operation capabilities between users and computer screens or designing interactive technologies based on spatial regions. In the case of non-vision, researchers can refer to the design suggestions in Table 4 when studying the space operation capabilities between users and computer screens or designing interactive technologies based on spatial regions.

Table 3. Design suggestions for selecting spatial targets under visual conditions.

	Longest Time	Shortest Time	Time Threshold	Highest Error Rate	Lowest Error Rate	Error Threshold	Proposal
Size	43.43 × 34.29	60.8 × 48	60.8 × 48	38 × 30	101.33 × 80	60.8 × 48	60.8 × 48
Region	D	B	/	C	D	/	D

Table 4. Design suggestions for selecting spatial targets under non-visual conditions.

	Longest Time	Shortest Time	Time Threshold	Highest Error Rate	Lowest Error Rate	Error Threshold	Proposal
Size	38 × 30	50.7 × 40	50.7 × 40	38 × 30	101.33 × 80	76 × 60	76 × 60
Region	D	B	/	C	B	/	D

Based on these results, we propose two techniques for two different application scenarios, described in the following paragraphs.

A spatial gesture recognition technique for surgery can help users select targets by using spatial region cognition and hand gestures during surgery. This technique is designed based on the partitioning strategies of a common operational spatial region array. This technique can meet the strict requirements of sanitary conditions during surgery (as opposed to a touchscreen and most other existing interfaces).

Non-visual selection is a system integrated with screen reading software allowing a visually impaired person to select targets easily. This technique is designed based on the partitioning strategies of a common operational spatial region array. Users can use this system to interact with the internet and web more easily. In addition, the user no longer needs a keyboard because this system uses Leap Motion to detect a users' hand motions and provides voice guidance when choosing targets and to do further work.

In the future, we will further expand the results of this study and contribute to technology accessibility for visually impaired individuals, including the exploration of a threshold for three-dimensional interaction.

Author Contributions: Methodology, validation, data curation, writing—review and editing, visualization, H.W.; conceptualization, formal analysis, resources, supervision, project administration, funding acquisition, J.Y.; software, data curation, methodology, writing—original draft preparation, Y.Z.; formal analysis, resources, supervision, project administration, investigation, Y.H.; investigation, supervision, methodology, resources, X.Z.; methodology, resources, formal analysis, supervision, S.W. All authors have read and agreed to the published version of the manuscript.

Funding: This work was funded by the National Natural Science Fund, grant (61741206).

Acknowledgments: The author thanks the editor and others for their comments and suggestions for this article.

Conflicts of Interest: The authors declare no conflict of interest.

References

1. Lee, M.; Bae, J. Deep learning based real-time recognition of dynamic finger gestures using a data glove. *IEEE Access* **2020**, *8*, 219923–219933. [CrossRef]
2. Marin, G.; Dominio, F.; Zanuttigh, P. Hand gesture recognition with leap motion and Kinect devices. In Proceedings of the 2014 IEEE International Conference on Image Processing (ICIP), Paris, France, 27–30 October 2014.
3. Chatterjee, I.; Xiao, R.; Harrison, C. Gaze+gesture: Expressive, precise and targeted free-space interactions. In Proceedings of the 2015 ACM on International Conference on Multimodal Interaction (ICMI '15), Association for Computing Machinery, New York, NY, USA, 9–13 November 2015; pp. 131–138.
4. Chalasani, T.; Smolic, A. Simultaneous segmentation and recognition: Towards more accurate ego gesture recognition. In Proceedings of the 2019 IEEE International Conference on Computer Vision Workshops, Seoul, Korea, 27–28 October 2019.
5. Martinez, J.; Griffiths, D.; Biscione, V.; Georgiou, O.; Carter, T. Touchless haptic feedback for supernatural VR Experiences. In Proceedings of the 25th IEEE Conference on Virtual Reality and 3D User Interfaces, VR 2018-Proceedings, Tuebingen/Reutlingen, Germany, 18–22 March 2018.
6. Park, J.; Jin, Y.; Cho, S.; Sung, Y.; Cho, K. Advanced machine learning for gesture learning and recognition based on intelligent big data of heterogeneous sensors. *Symmetry* **2019**, *11*, 929. [CrossRef]
7. Zhang, Y.; Yang, Z.; Zhang, G.; Wu, C.; Zhang, L. XGest: Enabling Cross-Label gesture recognition with RF signals. *ACM Trans Sens. Netw. TOSN* **2021**, *17*, 1–23. [CrossRef]
8. Liu, M.; Nancel, M.; Vogel, D. Gunslinger: Subtle arms-down mid-air interaction. In Proceedings of the 28th Annual ACM Symposium on User Interface Software & Technology (UIST '15), Association for Computing Machinery, New York, NY, USA, 9 August 2015; pp. 63–71.
9. Zaiţi, I.A.; Pentiuc, S.G.; Vatavu, R.D. On free-hand TV control: Experimental results on user-elicited gestures with leap motion. *Pers. Ubiquitous Comput.* **2015**, *5*, 821–838. [CrossRef]
10. Sa-Nguannarm, P.; Senavongse, W.; Charoenpong, T.; Kiatsoontorn, K. Hand movement recognition by using a touchless sensor for controlling images in operating room. In Proceedings of the 2018 International Electrical Engineering Congress (iEECON), Krabi, Thailand, 7–9 March 2018.
11. Liu, C. Leveraging physical human actions in large interaction spaces. In Proceedings of the Adjunct Publication of the 27th Annual ACM Symposium on User Interface software and Technology (UIST '14 Adjunct), Association for Computing Machinery, New York, NY, USA, 5–8 October 2014; pp. 9–12.
12. Aslan, I.; Krischkowsky, A.; Meschtscherjakov, A.; Wuchse, M.; Tscheligi, M. A leap for touch: Proximity sensitive touch targets in cars. In Proceedings of the 7th International Conference on Automotive User Interfaces and Interactive Vehicular Applications (AutomotiveUI '15), Association for Computing Machinery, New York, NY, USA, 1–3 September 2015; pp. 39–46.
13. Feng, J.; Shengping, Z.; Shen, W.; Yang, G.; Debin, Z. Multi-Layered gesture recognition with kinect. *J. Mach. Learn. Res.* **2015**, *1*, 227–254.
14. Spindler, M.; Stellmach, S.; Dachselt, R. PaperLens: Advanced magic lens interaction above the tabletop. In Proceedings of the ACM International Conference on Interactive Tabletops and Surfaces (ITS '09), Association for Computing Machinery, New York, NY, USA, 23–25 November 2009; pp. 69–76.
15. Cockburn, A.; Quinn, P.; Gutwin, C.; Ramos, G.; Looser, J. Air pointing: Design and evaluation of spatial target acquisition with and without visual feedback. *Int. J. Hum. Comput. Stud.* **2011**, *6*, 401–414. [CrossRef]
16. Gareth, Y.; Hamish, M.; Daniel, G.; Elliot, P.; Robert, B.; Orestis, G. Designing mid-air haptic gesture controlled user interfaces for cars. *ACM Hum.-Comput. Interact.* **2020**, *4*, 1–23.
17. Spindler, M.; Schuessler, M.; Martsch, M.; Dachselt, R. Pinch-Drag-Flick vs. spatial input: Rethinking zoom & pan on mobile displays. In Proceedings of the SIGCHI Conference on Human Factors in Computing Systems (CHI '14), Association for Computing Machinery, New York, NY, USA, 26 April–1 May 2014; pp. 1113–1122.
18. Jingyu, C.; Victoria, K.; Alan, M.; Elias, G.; David, H.; Joseph, T. A momentum-conserving implicit material point method for surface tension with contact angles and spatial gradients. *ACM Trans. Graph.* **2021**, *4*, 1–16.
19. Lee, M.; Kwahk, J.; Han, S.H.; Lee, H. Relative pointing interface: A gesture interaction method based on the ability to divide space. *Int. J. Ind. Ergon.* **2020**, *75*, 02878. [CrossRef]
20. Cha, Y.; Myung, R. Extended Fitts' law for 3D pointing tasks using 3D target arrangements. *Int. J. Ind. Ergon.* **2013**, *4*, 350–355. [CrossRef]

21. Brand, D.; Meschtscherjakov, A.; Büchele, K. Pointing at the HUD: Gesture interaction using a leap motion. In Proceedings of the 8th International Conference on Automotive User Interfaces and Interactive Vehicular Applications (AutomotiveUI '16 Adjunct), Association for Computing Machinery, New York, NY, USA, 24–26 October 2016; pp. 167–172.
22. Davis, M.M.; Gabbard, J.L.; Bowman, D.A.; Gracanin, D. Depth-Based 3D gesture multi-level radial menu for virtual object manipulation. In Proceedings of the 2016 IEEE Virtual Reality (VR), Greenville, SC, USA, 19–23 March 2016.
23. Schinazi, V.; Thrash, T.; Chebat, D. Spatial navigation by congenitally blind individuals. In *Wiley Interdisciplinary Reviews: Cognitive Science;* Wiley & Sons: New York, NY, USA, 2016.
24. Abidin, A.H.Z.; Xie, H.; Wong, K.W. Touch screen with audio feedback: Content analysis and the effect of spatial ability on blind people's sense of position of web pages. In Proceedings of the 2013 International Conference on Research and Innovation in Information Systems (ICRIIS), Kuala Lumpur, Malaysia, 27–28 November 2013.
25. Muniandy, M.; Sulaiman, S. Touch sensation as part of multimedia design elements to improve computer accessibility for the blind users. In Proceedings of the 2017 International Conference on Research and Innovation in Information Systems (ICRIIS), Langkawi, Malaysia, 16–17 July 2017.
26. Khambadkar, V.; Folmer, E. GIST: A gestural interface for remote nonvisual spatial perception. In Proceedings of the 26th Annual ACM Symposium on User Interface Software and Technology (UIST '13), Association for Computing Machinery, New York, NY, USA, 8–11 October 2013; pp. 301–310.
27. Khurana, R.; McIsaac, D.; Lockerman, E.; Mankoff, J. Nonvisual interaction techniques at the keyboard surface. In Proceedings of the 2018 CHI Conference on Human Factors in Computing Systems (CHI '18). Association for Computing Machinery, New York, NY, USA, 21–26 April 2018; pp. 1–12.
28. Wahab, M.; Mohamed, A.; Sukor, A.; Teng, O. Text reader for visually impaired person. In *Journal of Physics: Conference Series;* IOP Publishing: Bristol, UK, 2021.
29. Jinmo, K. VIVR: Presence of immersive interaction for visual Impairment virtual reality. *IEEE Access* **2020**, *8*, 196151–196159.

electronics

Article

The Study of Bending and Twisting Input Modalities in Deformable Interfaces

Jibin Yin [1], Shujie Bai [1], Yi Han [2], Xiangliang Zhang [3,*], Siyang Deng [1] and Shuoyu Wang [2]

1 Faculty of Information and Automation, Kunming University of Science and Technology, Kunming 650500, China; yinjibin@kust.edu.cn (J.Y.); whitesj@stu.kust.edu.cn (S.B.); siyangdeng@stu.kust.edu.cn (S.D.)
2 School of System Engineering, Kochi University of Technology, Kochi 780-8515, Japan; 258012g@gs.kochi-tech.ac.jp (Y.H.); Wang.shuoyu@kochi-tech.ac.jp (S.W.)
3 State Key Laboratory of Fluid Power and Mechatronic Systems, School of Mechanical Engineering, Zhejiang University, Hangzhou 310027, China
* Correspondence: xlzh@zju.edu.cn

Abstract: The deformable input provides users with the ability of physical operation equipment to interact with the system. In order to facilitate further development in flexible display interactive technology, we devised FlexSheet, an input device that can simulate the deformation environment. This paper presents two forms of deformation input, bending and twisting, with regard to three selection techniques. We conduct a controlled experiment to select discrete targets by combining two input forms and three selection strategies, taking into account the influence of visual feedback. Further, we use the deformation angle to reflect the degree of deformation and map it to the experimental variables. In accordance with the experimental results, we analyze the experimental performance under three evaluation indexes and prove the viability of our selection technology in bending and twisting input modes. Finally, we provide suggestions on the control level in bending and twisting input modes, respectively.

Keywords: flexible display; bending input; twisting input; selection technology

Citation: Yin, J.; Bai, S.; Han, Y.; Zhang, X.; Deng, S.; Wang, S. The Study of Bending and Twisting Input Modalities in Deformable Interfaces. *Electronics* **2021**, *10*, 2991. https://doi.org/10.3390/electronics10232991

Academic Editor: J.-C. Chiao

Received: 28 October 2021
Accepted: 28 November 2021
Published: 1 December 2021

1. Introduction

General displays have achieved good performance after technical iterations, like liquid crystal display (LCD), which have been widely used in everyday life. Limited to its own structural characteristics, it can only express two-dimensional information and the degree of freedom of interaction is limited to the screen surface [1]. With the introduction of the concept of organic user interfaces (OUIs) [2], the research on deformable user interfaces (DUI) [3,4] and its interaction technology is becoming increasingly extensive. The interactive feature of flexible devices is that we can change their shape in accordance with different functions, such as using deformation as input parameters to open music [5,6], manipulate maps [7], and even prove that it can be used to input passwords, which enhances interactive security during the process of innovation [8], thereby providing an advantage that traditional screen technology does not have. This additional input mode not only greatly increases the bandwidth of human-computer interaction [9] but also inspires researchers to explore greater possibilities of this interaction technology.

The design of the deformation simulation prototype is the premise and foundation of this kind of researches, and the main objective is to simulate the input environment of deformation equipment. Certain researchers use electrical components as deformation sensors [5,8,10–12]. The principle is that when electrical components are deformed, their voltage changes accordingly. Moreover, different degrees of deformation can be simulated by tracking the reflective ball attached to the surface [7,13]. There are some other related work, such as the use of optical sensors [14,15] and the use of fiber-based deformation sensor clusters [16] to detect human intentional deformation, etc. Our implementation scheme is based on the OptiTrack motion capture system. In this system, multiple infrared cameras collect reflective identification points in space in real time and calculate the intersection of multiple direction vectors after completing target correction and recognition in order to determine the world coordinates. In addition, this infrared optical positioning technology can effectively eliminate visible light interference, improve system robustness, and yield a high data transmission rate, which is similar to what has been found in extant research [17].

There are numerous studies on enhancing interactive expressiveness based on prototypes. The works of deformation gesture has attracted the interest of many researchers. The remarkable feature of gesture operation is that it can give users a more intuitive experience and a strong sense of participation in the interaction process. Some researchers directly bind specially designed deformable gestures with functional requirements to improve interactive bandwidth, such as controlling smart TV [18]. A larger trend is that researchers collect user-defined deformation gesture consensus sets [19,20] to understand users' natural interaction habits and solve some practical problems.

In the process of exploring deformation gestures, researchers found that devices using bending gestures are the most effective for rapid response to continuous bipolar variables [21], and suggested to maintain two orders of magnitude [22]. If it is a one handed operation, the deformation gesture in the upper right corner is a better choice [23].

In addition to taking the deformation degree of the equipment directly as the input parameter [5,7,17], many researchers have also discussed the input potential of the combination of deformation and other input methods, such as pressure [24,25], touch [26,27], and even provide new interaction methods for people with visual impairment by combining hearing [28] and kinesthetic [27].

Certain researchers believe that the advantage of deformation input is not only a single command binding. In other words, this interactive technology has a vivid metaphor, which is related to the user's behavior and ideas in a few cases. For example, by reading [13], we can simulate the feeling of actual physical paper on the deformation device. When we need to turn the page [10], it is like using real documents. Scenes in virtual games—such as fishing, archery, or golf games are similar examples [11], and embodied in intelligent wearable healthcare equipment [29].

In this study, we design a prototype called FlexSheet that supports deformation as input and based on the OptiTrack motion capture system. This prototype can provide the interactive characteristics of flexible devices to simulate a deformable input environment (Figure 1). We call this the deformation controller. The deformation controller consists of two thin and soft polyvinyl chloride (PVC) parts of size 20 cm $\times$ 20 cm on which the reflective traceable ball is fixed. We use two input forms: bending and twisting. In addition, two visual feedback mechanisms are set up—full vision (FV) and partial vision (PV). Simultaneously, we use an angle to describe the degree of deformation. With the increase in force feedback [20,30], the angle of deformation becomes increasingly larger. For the quantitative analysis of deformation, we mapped the angle to the experimental variables and took the maximum bending angle of 180° in the horizontal state. Finally, through a simple input form, we checked the user's ability to control deformation input in order to obtain relevant design guidelines.

The remainder of the paper is organized in the following manner. Section 2 introduces the experimental materials, including input mode, visual feedback, and selection technologies. Section 3 describes controlled experiments used to study deformation input performance. Section 4 analyzes the experimental results from the perspective of three indexes: completion time (MT), error rate (ER), and crossing times (NC); thereafter the best order of magnitude suggestions under the two input modes are provided, respectively. Section 5 discusses and summarizes the paper.

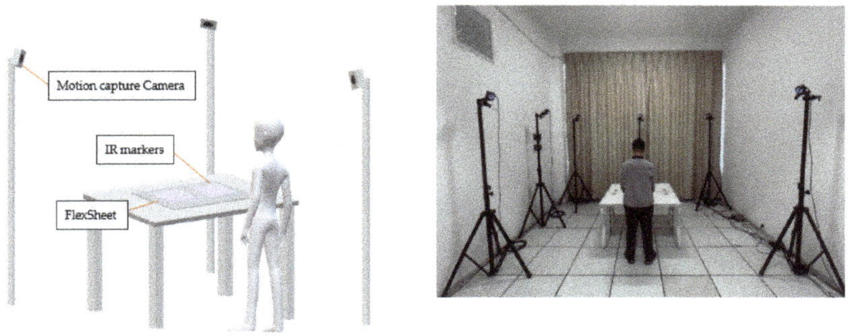

(a) (b)

Figure 1. Composition and operation of deformation input: (**a**) Deformation input environment diagram; (**b**) Experiment with OptiTrack and FlexSheet.

2. Materials Instruction

2.1. Input Modes

Most flexible screen studies have demonstrated that bending and twisting are two basic types in the design of deformation gestures. Combined with our research content, we provide the following design guidelines for deformation gestures:

1. The deformation gestures must be universal (that is, they can be realized on as many materials as possible), because different materials allow different forms of deformation; plastic sheets are the most similar type of flexible display material that is currently available [18];
2. The deformation gestures must give the feeling of using real thing [19];
3. The deformation gestures must have a high degree of consistency in orientation [10];
4. The deformation gestures must have powerful powerful metaphors [19].

According to the above design principles, we examined two kinds of deformation gestures: bending and twisting. The basic operation example is depicted in Figure 2.

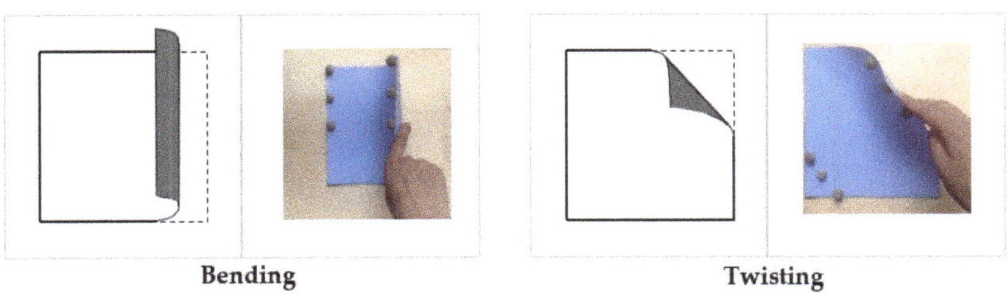

(a) (b)

Figure 2. Two deformation input models: (**a**) bending; (**b**) twisting.

2.2. Visual Feedback

As depicted in Figure 3, we used two visual feedback conditions: FV and PV. Under FV conditions the cursor is always visible, as illustrated in Figure 3a,b. Under PV conditions, the cursor is only displayed at the initial state of the selection task and is hidden once the selection task is initiated, as depicted in Figure 3c,d. Under PV conditions, the user must depend on previous experience in FV conditions to complete target selection. This simulates a situation in which expert users may learn to use deformation input intuitively and without excessive cursor support or visual cues, which is similar to the behavior that expert users display when tagging menus [31].

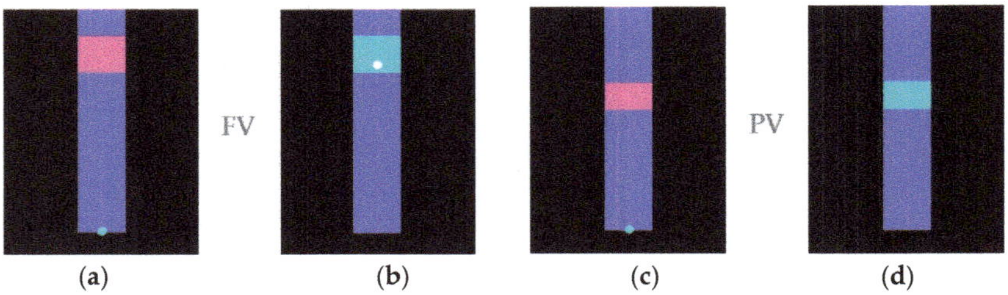

Figure 3. Target selection experiment interface: (**a**) The initial state with full visual feedback; (**b**) The target selected state with full visual feedback; (**c**) The initial state with partial visual feedback; (**d**) The target selected state with partial visual feedback.

2.3. Selection Technologies

Once the cursor enters the target, the user must have a mechanism to identify and select options in the general graphical user interface (GUI), as general determination measures are usually performed by clicking the buttons on the mouse. Consequently, we tested three alternative technological solutions to replace the traditional mouse functions:

1. Click: Move the cursor by bending or twisting the FlexSheet. When the cursor appears within the target rectangle, click the task button to complete the target selection;
2. Dwell: Bending or twisting the FlexSheet to move the cursor until it appears within the target, then keep the cursor in the target for a specified period of time to complete the target selection (in our experiment, a delay of 0.7 seconds was used);
3. Quick Release: Move the cursor by bending or twisting the FlexSheet. When the cursor appears in the target, quickly remove the FlexSheet from the deformed state.

3. Controlled Experiment

Through this experiment, our objective was to study the human ability to complete the task of selecting discrete string targets when using deformation inputs. This mainly includes the influence of visual feedback on the selection process and selection results, actual performance of three target selection strategies as well as the control level under different input modes.

3.1. Participants

We recruited 12 participants, including ten males and two females, aged 23–31 years, with an average age of 26 years. All participants are right-handed and have some user graphical interface experience, but have no experience of using deformation input.

3.2. Apparatus and Environment

The experiment was conducted on Lenovo computers running Windows10, 3.8 GHz, screen size of 27 inches and resolution of 1920 by 1080 pixels. The experiment also used a motion capture device, OptiTrack, which is composed of eight cameras with Prime13/13W specification and connected through Cisco switch. The deformable plate is made of a plastic sheet (20 cm $\times$ 20 cm), and markers accessories is fixed to the prototype. The products in the experiment were compiled by the NatNetSDK3.0.1 provided by OptiTrack official website and run in the Visual Studio 2015 environment.

3.3. Task

Our long-term goal is to explore the general design criteria for deformation interactions, with a particular emphasis on developing design guidelines for deformation interactions based on bending and twisting modes.

We used a serial target selection task. Control the cursor to move vertically by deforming the FlexSheet. The deformation value of 180° is uniformly mapped to 256 pixels. Within the distance of 256 pixels, we draw a set of continuous rectangular targets. In each experiment, a target rectangle to be selected will be highlighted in purple. The user's task is to apply appropriate deformation and move the cursor to the target rectangle with different distances and different widths. When the cursor enters the range of the target rectangle, the target rectangle changes from purple to green to give visual feedback to the subject.

It should be noted that the cursor is displayed in the initial state of the task. Since we provide two different visual feedback mechanisms, when the cursor starts to move, the full visual feedback supports the full display of the cursor, while the cursor under partial visual feedback will be hidden.

After the target to be selected gives color feedback, the subject needs to use the three selection technologies we provide to quickly complete the target selection, and the experimental task is over. Subjects will then repeat this task many times under different combinations of conditions.

3.4. Performance Measures

MT is the dependent variable of the experimental task, which represents the time it takes to select each target, defined as the time (milliseconds) from moving the cursor after the target appears until the target is selected; ER means the number of incorrect selections in the target selection task accounts for the total number of selections the ratio; NC represents the cumulative number of crossovers that enter the target and then leave the target when selecting each target. The functions of these evaluation indicators complement each other. ER and MT provide us with an overview of the overall success rate, and NC provides us with an indication of the degree of control of the user's application of deformed input.

3.5. Procedure and Design

A within-subjects full factorial design with repeated measures was used. The independent variables were selection modes (Bending and Twisting), selection technologies (Click, Dwell and Quick Release), visual feedback condition (FV, PV), the distance from the starting point to the target (D = 37, 72, 109, 144) and the target width (W = 15, 18, 22.5, 30, 45). The units of distance and width are represented by the deformation angle.

For each visual feedback, each participant will be exposed to two input modes. In each input mode, participants need to complete the experiments of all three selection technologies. For each selection technology, 20 D-W combination conditions are included and presented in random order. The experiment order was counterbalanced among participants using a Latin Square design.Each participant repeated the entire task three times to complete a total of 8640 individual tasks. In summary, the experiment consisted of:

- Selection modes (Bending and Twisting);
- Visual feedback conditions (FV and PV);
- Selection technologies (Click, Dwell and Quick Release);
- Target distance (D = 37, 72, 109, 144);
- Target width (W = 15, 18, 22.5, 30, 45);
- 12 Participants
- 3 Repetitions

Before the experiment, we explained the task, and performed a brief demonstration. Participants were given a short warm-up set of trials to familiarize themselves with input modes, visual feedback and selection technologies. Participants were instructed to perform the task as quickly and accurately as possible. Participants can take a short break during the experiment. The experiment lasted approximately 1 hour for each participant. To analyze the experimental results, we recorded the relevant parameters of each target selection task, including input modes, target size, target distance, time spent, errors, and crossing times.

4. Results

We analyzed the collected measures by conducting a repeated measures ANOVA ($\alpha = 0.05$) that used the three-interaction technique and two-visual feedback condition as independent variables. Further, we used Greenhouse-Geisser corrected values in the analysis when the assumption of sphericity was violated (tested with Mauchly's test). Pairwise t-tests with Bonferroni corrections were used for post hoc tests. The test was considered an outlier if the time required to complete the task exceeded two standard deviations of the average completion time of the task. A total of 282 outliers were eliminated, which accounted for 3.3% of the collected data.

4.1. Time

As depicted in Figure 4, in bending mode, we found that different selection techniques had no significant effect on performance time ($F_{1.342,14.764} = 2.504$, $p = 0.129$). Further, we found no significant interaction between interaction technique and visual feedback ($F_{2,22} = 3.286$, $p = 0.056$), as shown in Figure 5.

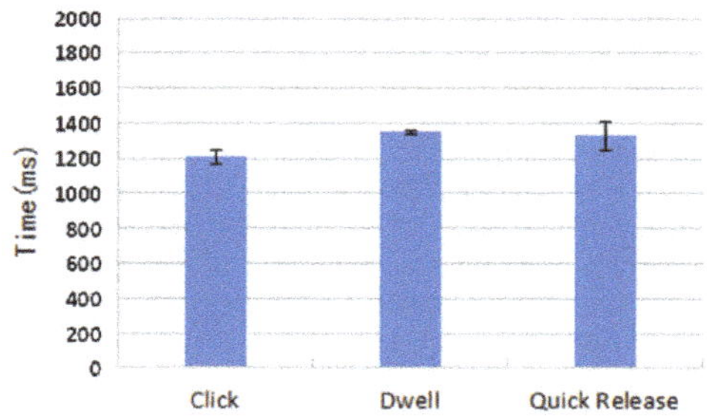

Figure 4. Elapsed time with diverse selection techniques in bending mode (error bars indicate 95% confidence intervals).

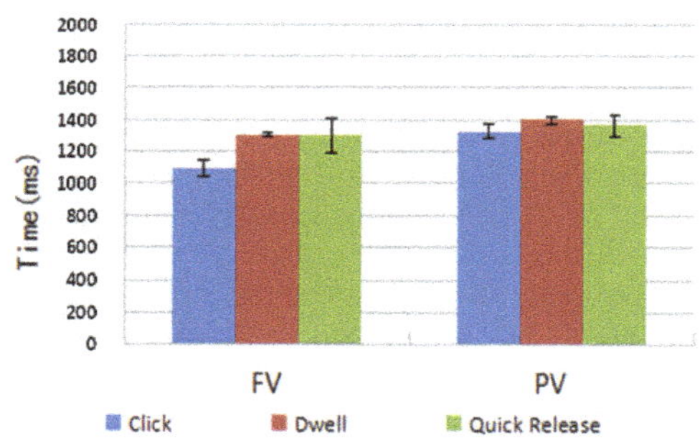

Figure 5. Elapsed time for each selection technique under diverse visual feedback in bending mode (error bars indicate 95% confidence intervals).

Although there was no significant difference in the performance of time between the three interactive technologies, it can be found that the most time was spent in Dwell, followed by Quick Release, and the then Click.

As illustrated in Figure 6, in twisting mode, we found that different selection techniques had no significant effect on performance time ($F_{1.165,12.818} = 2.284$, $p = 0.154$). As depicted in Figure 7, we found a significant interaction between technology and visual feedback ($F_{1.171,12.882} = 4.749$, $p = 0.044$). The visual feedback condition has a significant effect on Click time ($p = 0.002$) as well as on Dwell and Quick Release times ($p < 0.001$, $p = 0.002$). The overall trend is that the Quick Release time under PV is significantly higher than that under FV ($p = 0.004$, $p = 0.034$); there was no significant difference among all technologies ($p > 0.1$).

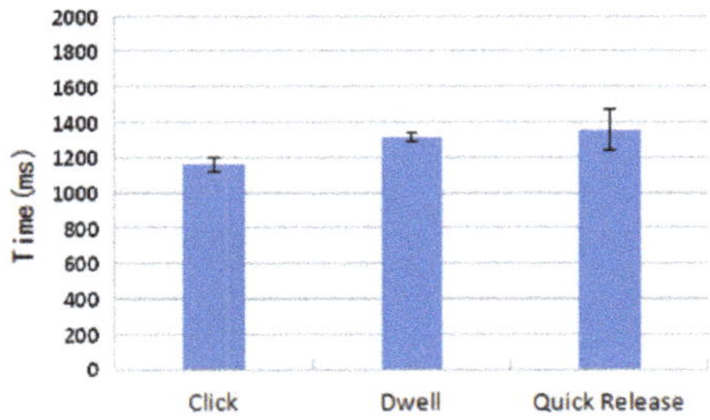

Figure 6. Elapsed time with diverse selection techniques in twisting mode (error bars indicate 95% confidence intervals).

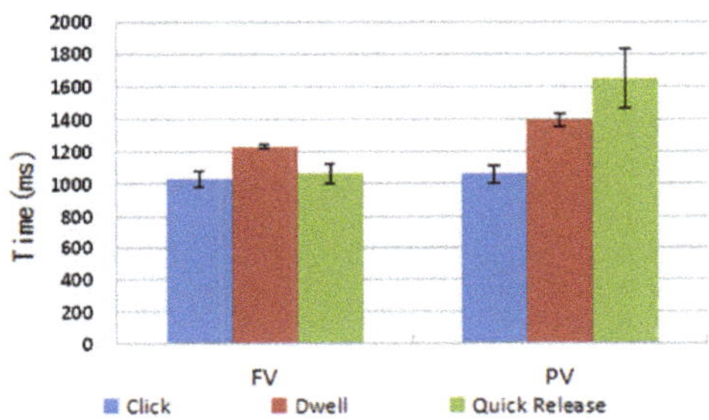

Figure 7. Elapsed time for each selection technique under diverse visual feedback in twisting mode (error bars indicate 95% confidence intervals).

4.2. Accuracy

In bending mode, we found that selection technology has a significant effect on the accuracy ($F_{1.243,13.673} = 7.9$, $p = 0.01$), as illustrated in Figure 8. The post hoc tests revealed no significant differences among the technologies except Click and Quick Release ($p = 0.021$) ($p > 0.07$).

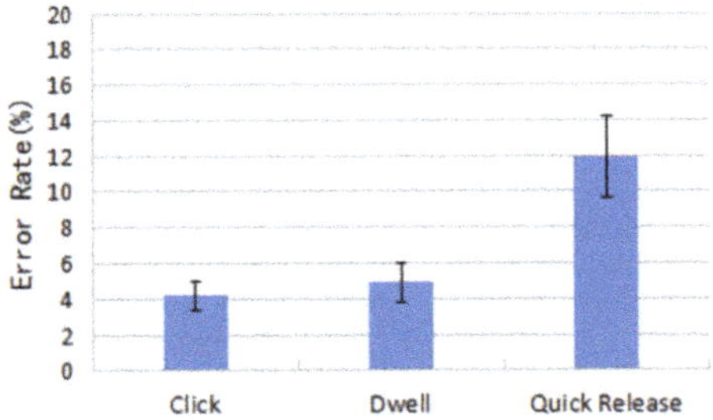

Figure 8. Error rate with diverse selection techniques in bending mode (error bars indicate 95% confidence intervals).

As illustrated in Figure 9, there was a significant interaction between selection techniques and visual feedback ($F_{2,22} = 3.609$, $p = 0.044$).

The level of visual feedback had no obvious effect on the error rate for Click ($p = 0.604$); however, a significant effect was found on the error rates for Dwell and Quick Release ($p = 0.006$, $p = 0.005$). The general trend was that the error under PV was significantly higher than that under FV. The post hoc tests revealed that under FV conditions, there were no significant differences between technologies; under PV conditions ($p > 0.1$), Click and

Quick Release were significant (p = 0.012). In addition, there are no significant differences among other technologies (p > 0.1).

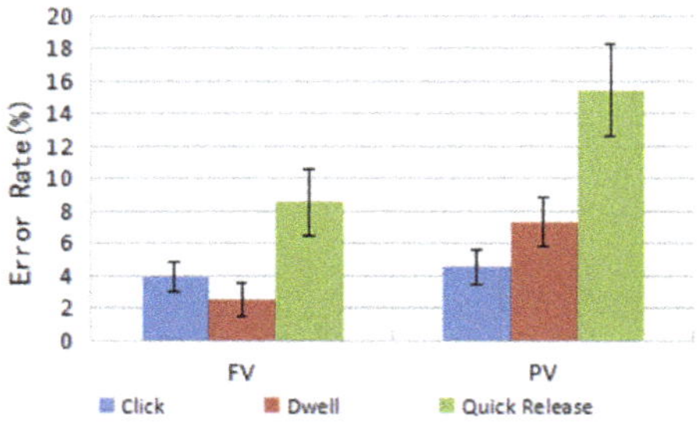

Figure 9. Error rate for each selection technique under diverse visual feedback types in bending mode (error bars indicate 95% confidence intervals).

As depicted in Figure 10, in twisting mode, we found a main effect on the error rate of different selection techniques ($F_{1.167,12.839}$ = 11.812, p < 0.001).

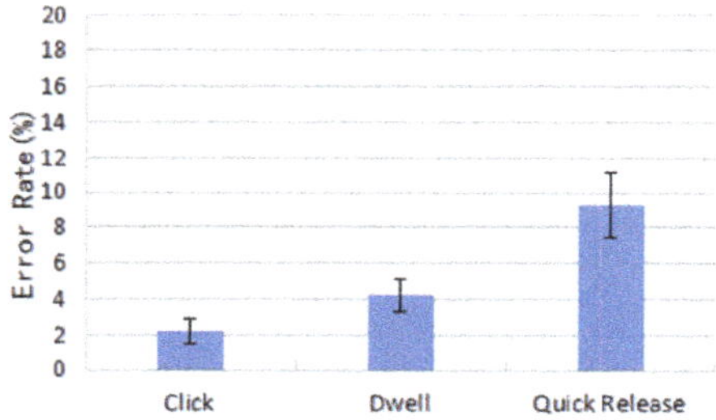

Figure 10. Error rate with different selection techniques in twisting mode (error bars indicate 95% confidence intervals).

The post hoc tests showed significant differences among all selection technologies (p = 0.02, p = 0.00, p = 0.047). Further, as illustrated in Figure 11, we found no significant effect between selection techniques and visual feedback ($F_{1.305,14.35}$ = 1.406, p > 0.2).

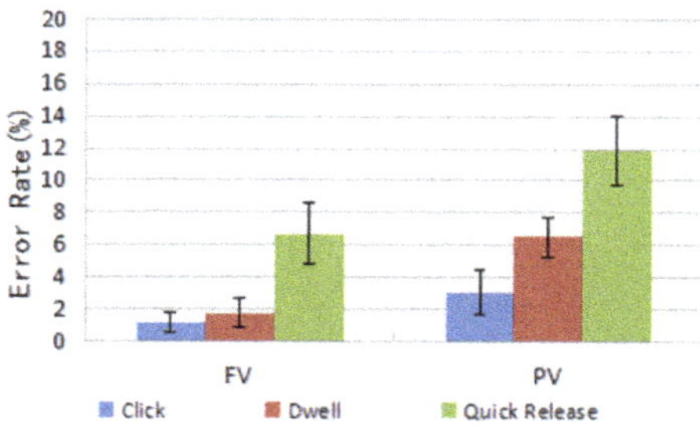

Figure 11. Error rate for each selection technique under diverse visual feedback in twisting mode (error bars indicate 95% confidence intervals).

4.3. Indication of Control

As illustrated in Figure 12, in bending mode, we found that this technique had a significant impact on the number of performance crosses ($F_{1.254,13.791}$ = 8.527, p = 0.008). The post hoc tests revealed that there was no significant difference among all technologies, except Dwell and Quick Release (p = 0.022) (p > 0.05).

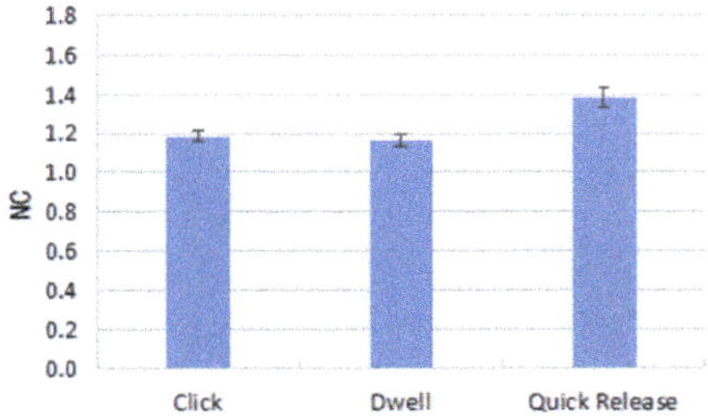

Figure 12. NC rate with diverse techniques in bending mode (error bars indicate 95% confidence intervals).

As presented in Figure 13, there was no significant interaction between technology and visual feedback ($F_{1.769,19.464} = 2.771$, $p = 0.092$).

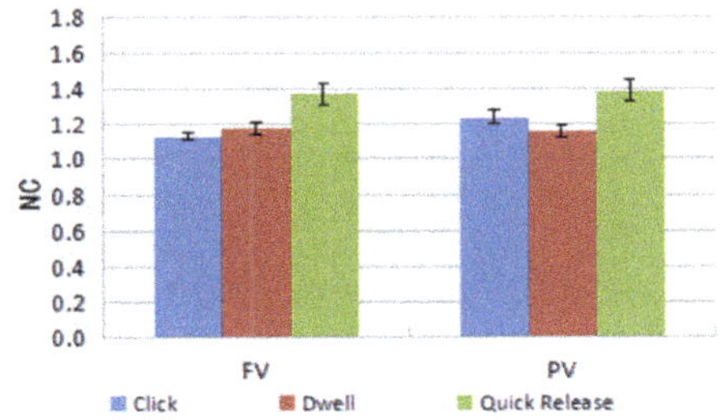

Figure 13. NC rate for each technique under diverse visual feedback in bending mode (error bars indicate 95% confidence intervals).

As depicted in Figure 14, in twisting mode, we found that different techniques had a significant effect on the NC ($F_{1.353,14.885} = 17.716$, $p < 0.001$).

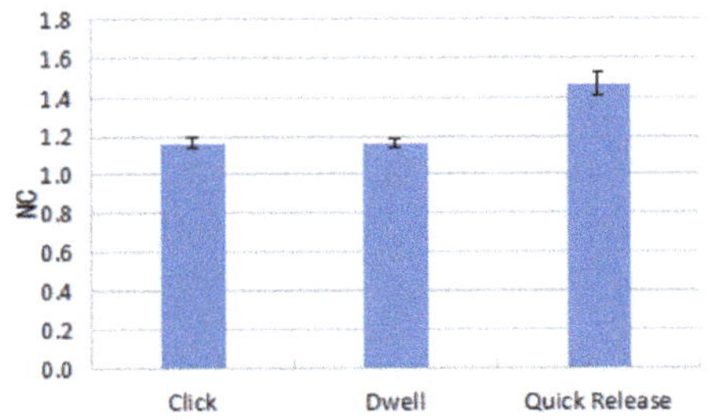

Figure 14. NC rate with diverse techniques in twisting mode (error bars indicate 95% confidence intervals).

There was no significant difference between Click and Dwell ($p = 0.032$), as revealed by the post hoc tests, but there was no significant difference between Click and Dwell ($p = 0.066$, $p = 0.069$).

As depicted in Figure 15, there was no significant interaction between technology and visual feedback ($F_{1.218,13.397} = 5.783$, $p = 0.026$). In addition, the level of visual feedback has no significant effect on the NC of Click ($p = 0.084$), Dwell ($p = 0.169$), and Quick Release ($p = 0.009$). The general trend was that the NC under PV condition is significantly higher than that under FV condition. In FV condition, Click and Quick Release ($p = 0.011$), Dwell and Quick Release ($p = 0.007$) differed significantly among different technologies, except

Click and Dwell (p = 1). Under PV conditions, Click and Quick Release (p = 0.01) differed significantly from Dwell and Quick Release (p = 0.007), except for Click and Dwell (p = 1).

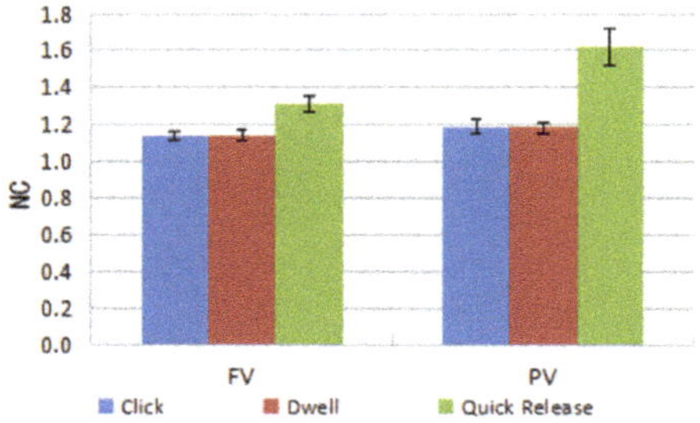

Figure 15. NC rate for each technique under diverse visual feedback in twisting mode (error bars indicate 95% confidence intervals).

4.4. Optimum Number of Deformation Levels

One of the intentions of this research is to ascertain the number of independent bending levels (nLevels) a user can distinguish under a given level of device performance. Except the Quick Release, when nLevels ≤ 8, all selection technology error values between 1–2%, and there was no significant difference in the error rate (Figure 16). Except Quick Release, when nLevels ≤ 6, the NC of FV was stable around 1.1 in each selection technology; moreover, all the selection technologies of NC had no significant differences (Figure 17). In bending mode, these results indicate that six levels of deformation were a reasonable number in bending mode.

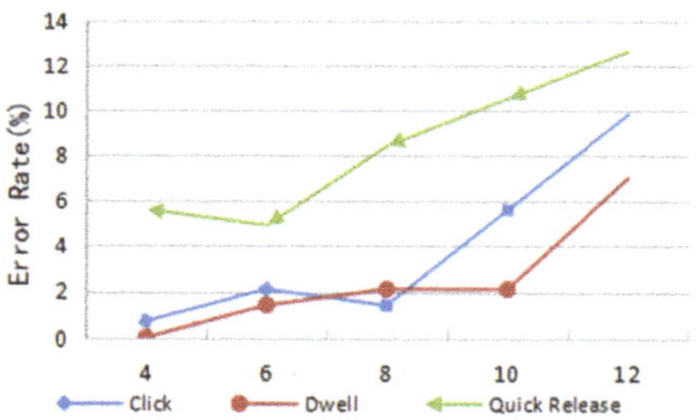

Figure 16. ER with nLevels under FV condition (bending mode).

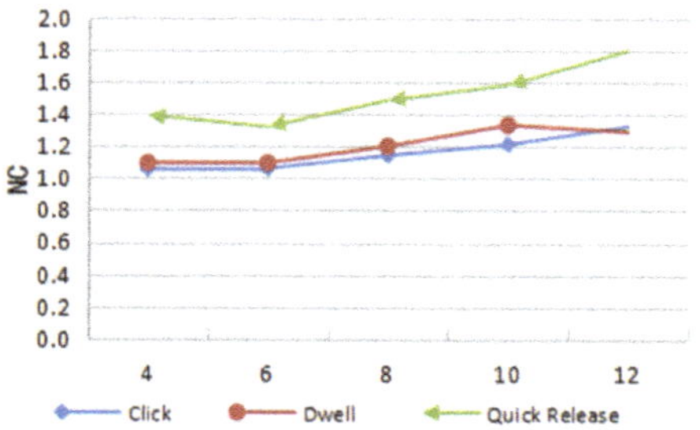

Figure 17. NC with nLevels under FV condition (bending mode).

Except the Quick Release, when nLevels ≤ 8 and when all selection technology error values are between 1% and 3%, there were no significant differences between the error rates of the technology selected (Figure 18). In addition to Quick Release, when nLevels ≤ 8, of all stable technologies of NC in 1.1%, there was no significant difference (Figure 19). In twisting mode, these results indicate that eight levels of deformation were a reasonable number for humans to comfortably control in twisting mode.

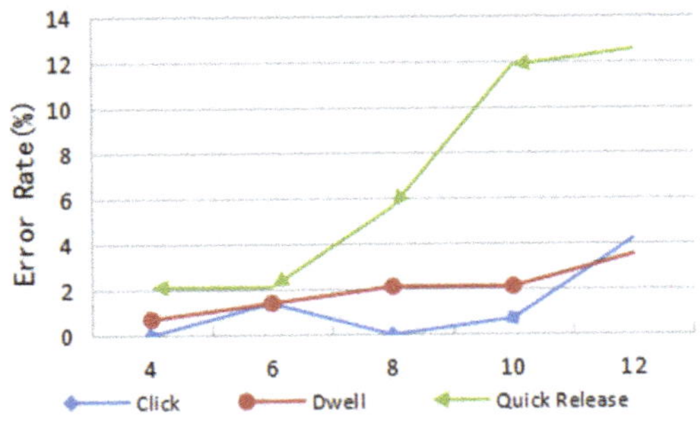

Figure 18. ER with nLevels under the FV condition (twisting mode).

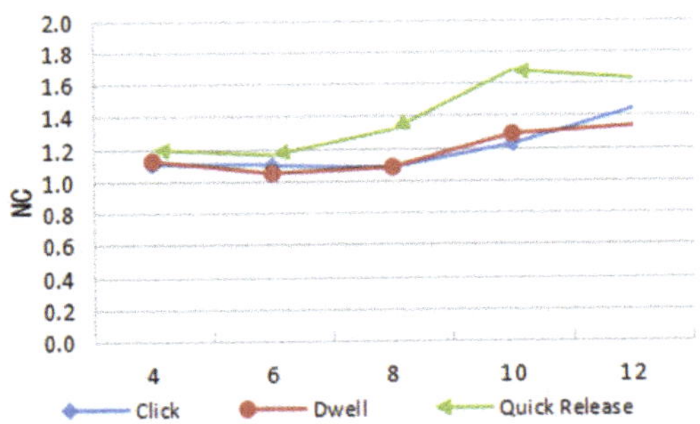

Figure 19. NC with nLevels under the FV condition (twisting mode).

5. Discussion and Conclusions

In this study, we examined the interaction technology of a flexible screen and designed a simulation prototype called FlexSheet, which can support deformation inputs. Among with the OptiTrack system, a deformation input environment was constructed. In this context, we summarized two basic deformation models: bending and twisting. In order to study the ability of humans to control deformation, we introduced the deformation angle as a reflection of the degree of deformation. Further, we designed a discrete object selection experiment using three selection techniques (Click, Dwell, and Quick Release) and two types of visual feedback FV and PV to simulate novice and expert users. According to the experimental performance of participants, a series of performance indicators were analyzed in detail. The results revealed that in two input modes, less time was spent on simple Click than Dwell and Quick Release, a finding that is in keeping with our hypothesis. The main reason is that the Click selection is more in line with a user's operating habits. Another reason is that the other two selection methods need to be implemented with a certain time delay, thereby reducing the selection efficiency. In terms of accuracy, both models had similar performance, with the lowest error rate for Click, followed by Dwell, and then Quick Release. In terms of stability (NC), from the input mode, the number of twisting intersections was lesser than that of bending intersections, but the overall performance is similar. From the perspective of different selection technologies, the average number of intersections to Dwell is the lowest, followed by Click, and then Quick Release. This is consistent with the feedback of participants that rapid release is more difficult to control than other technologies, and it is easier to remain within the target range using bending input, which leads to higher time cost, error rate, and crossing times. According to our measurement and analysis, the maximum deformation grade can be divided into six levels in bending and eight levels in twisting. The division basis of deformation levels comes from the mapping of deformation angle to width variables in the task, and the division results are reflected by user feedback. It is worth noting that in each index analysis, the overall performance of FV is better than that of PV.

Further, although we discussed the specific performance of the two deformation input modes, respectively, we believe that the research on the combination of the two input modes has good potential in future research on flexible screen technology. In addition, we believe that the research on ergonomic models based on deformation input is of great significance. We are optimistic regarding the future of flexible interactive technology, and our study can provide designers a few useful guidelines for designing interactive techniques based on flexible screen.

Author Contributions: Conceptualization, funding acquisition, project administration, supervision, J.Y.; methodology, data analysis, experimental verification, writing, S.B.; investigation, supervision, Y.H.; investigation, supervision, X.Z.; software, data curation, S.D.; investigation, supervision, S.W. All authors have read and agreed to the published version of the manuscript.

Funding: This research was supported by the National Natural Science Foundation of China (61741206).

Acknowledgments: The authors would like to thank the editor and anonymous reviewers for their useful comments for improving the quality of this paper.

Conflicts of Interest: The authors declare no conflict of interest.

References

1. Inoue, Y.; Itoh, Y.; Onoye, T. TuVe: A flexible display with a tube. In *SIGGRAPH Asia 2018 Emerging Technologies (SA'18)*; Association for Computing Machinery: New York, NY, USA, 2018; pp. 1–2.
2. David, H.; Roel, V. Organic user interfaces: Designing computers in any way, shape, or form. *ACM* **2008**, *51*, 48–55.
3. Mone, G. The future is flexible displays. *Commun. ACM* **2013**, *56*, 16–17. [CrossRef]
4. Girouard, A.; Eady, A.K. Deformable User Interfaces: Using Flexible Electronics for Human mputer Interaction. In Proceedings of the International Flexible Electronics Technology Conference, Ottawa, ON, Canada, 7–9 August 2018. [CrossRef]
5. Lahey, B.; Girouard, A.; Burleson, W.; Vertegaal, R. PaperPhone: Understanding the Use of Bend Gestures in Mobile Devices with Flexible Electronic Paper Displays. In Proceedings of the SIGCHI Conference on Human Factors in Computing Systems (CHI'11), New York, NY, USA, 5–10 May 2011; pp. 1303–1312. [CrossRef]
6. Gomes, A.; Priyadarshana, L.; Carrascal, J.P.; Vertegaal, R. WhammyPhone: Exploring Tangible Audio Manipulation Using Bend Input on a Flexible Smartphone. In Proceedings of the 29th Annual Symposium on User Interface Software and Technology (UIST'16 Adjunct), Tokyo, Japan, 16–19 October 2016; pp. 159–161.
7. Gallant, D.T.; Seniuk, A.G.; Vertegaal, R. Towards more paper-like input: Flexible input devices for foldable interaction styles. In Proceedings of the 21st Annual ACM Symposium on User Interface Software and Technology (UIST'08), New York, NY, USA, 19–22 October 2008; pp. 283–286. [CrossRef]
8. Maqsood, S.; Chiasson, S.; Girouard, A. Bend Passwords: Using gestures to authenticate on flexible devices. *Pers. Ubiquitous Comput.* **2016**, *20*, 573–600. [CrossRef]
9. Ramos, G.; Boulos, M.; Balakrishnan, R. Pressure widgets. In Proceedings of the SIGCHI Conference on Human Factors in Computing Systems, New York, NY, USA, 24–29 April 2004; pp. 487–494. [CrossRef]
10. Watanabe, J.I.; Mochizuki, A.; Horry, Y. Bookisheet: Bendable device for browsing content using the metaphor of leafing through the pages. In Proceedings of the 10th International Conference on Ubiquitous Computing (UbiComp'08), New York, NY, USA, 21–24 September 2008; pp. 360–369. [CrossRef]
11. Ye, Z.; Khalid, H. Cobra: Flexible displays for mobilegaming scenarios. In Proceedings of the CHI'10 Extended Abstracts on Human Factors in Computing Systems (CHI EA'10), New York, NY, USA, 10–15 April 2010; pp. 4363–4368. [CrossRef]
12. Schwesig, C.; Poupyrev, I.; Mori, E. Gummi: A bendable computer. In Proceedings of the SIGCHI Conference on Human Factors in Computing Systems (CHI'04), New York, NY, USA, 24–29 April 2004; pp. 263–270. [CrossRef]
13. Holman, D. Paper windows: Interaction techniques for digital paper. In Proceedings of the SIGCHI Conference on Human Factors in Computing Systems (CHI'05), New York, NY, USA, 2–7 April 2005; pp. 591–599.
14. Weigel, M.; Steimle, J. DeformWear: Deformation Input on Tiny Wearable Devices. *Proc. ACM Interact. Mob. Wearable Ubiquitous Technol.* **2017**, *1*, 23. [CrossRef]

15. Hosono, S.; Nishimura, S.; Iwasaki, K.; Tamaki, E. Gesture Recognition System using Optical Muscle Deformation Sensors. In Proceedings of the 2019 2nd International Conference on Electronics, Communications and Control Engineering (ICECC 2019), Phuket, Thailand, 13–16 April 2019; pp. 12–15.
16. Fellion, N.; Pietrzak, T.; Girouard, A. FlexStylus: Leveraging Bend Input for Pen Interaction. In Proceedings of the 30th Annual ACM Symposium on User Interface Software and Technology (UIST'17), Quebec City, QC, Canada, 22–25 October 2017; pp. 375–385.
17. Herkenrath, G.; Karrer, T.; Borchers, J. Twend: Twisting and bending as new interaction gesture in mobile devices. In Proceedings of the Extended Abstracts on Human Factors in Computing Systems (CHI'08), New York, NY, USA, 5–10 April 2008; pp. 3819–3824. [CrossRef]
18. Lee, S.S.; Maeng, S.; Kim, D.; Lee, K.P.; Lee, W.; Kim, S.; Jung, S. FlexRemote: Exploring the Effectiveness of Deformable User Interface as an Input Device for TV. In *HCI International 2011—Posters' Extended Abstracts. HCI 2011*; Stephanidis, C., Ed.; Communications in Computer and Information Science; Springer: Heidelberg/Berlin, Germany, 2011; Volume 174. [CrossRef]
19. Lee, S.S.; Kim, S.; Jin, B.; Choi, E.; Kim, B.; Jia, X.; Lee, K.P. How users manipulate deformable displays as input devices. In Proceedings of the SIGCHI Conference on Human Factors in Computing Systems (CHI'10), New York, NY, USA, 10–15 April 2010; pp. 1647–1656.
20. Borah, P.P.; Sorathia, K. Natural and Intuitive Deformation Gestures for One-handed Landscape Mode Interaction. In Proceedings of the Thirteenth International Conference on Tangible, Embedded, and Embodied Interaction (TEI '19), Tempe, AZ, USA, 17–20 March 2019; pp. 229–236. [CrossRef]
21. Ahmaniemi, T.T.; Kildal, J.; Haveri, M. What is a device bend gesture really good for? In Proceedings of the SIGCHI Conference on Human Factors in Computing Systems (CHI'14), New York, NY, USA, 26 April–1 May 2014; pp. 3503–3512.
22. Warren, K.; Lo, J.; Vadgama, V.; Girouard, A. Bending the Rules: Bend Gesture Classification for Flexible Displays. In Proceedings of the SIGCHI Conference on Human Factors in Computing Systems (CHI'13), New York, NY, USA, 27 April–2 May 2013; pp. 607–610. [CrossRef]
23. Girouard, A.; Lo, J.; Riyadh, M.; Daliri, F.; Eady, A.K.; Pasquero, J. One-Handed Bend Interactions with Deformable Smartphones. In Proceedings of the 33rd Annual ACM Conference on Human Factors in Computing Systems, New York, NY, USA, 18–23 April 2015; pp. 1509–1518.
24. Gotsch, D.; Zhang, X.; Carrascal, J.P.; Vertegaal, R. HoloFlex: A Flexible Light-Field Smartphone with a Microlens Array and a P-OLED Touchscreen. In Proceedings of the 29th Annual Symposium on User Interface Software and Technology (UIST'16), New York, NY, USA, 16–19 October 2016; pp. 69–79. [CrossRef]
25. Ansara, R.; Girouard, A. Augmenting bend gestures with pressure zones on flexible displays. In Proceedings of the 16th International Conference on Human-Computer Interaction with Mobile Devices & Services (MobileHCI'14), New York, NY, USA, 24–27 September 2014; pp. 531–536. [CrossRef]
26. Kildal, J.; Lucero, A.; Boberg, M. Twisting touch: Combining deformation and touch as input within the same interaction cycle on handheld devices. In Proceedings of the 15th International Conference on Human-Computer Interaction with Mobile Devices and Services (MobileHCI'13), New York, NY, USA, 27–30 August 2013; pp. 237–246.
27. Borah, P.P. Deformation Gesture-based Input Method for Non-visual Primitive Geometric Shape Drawing. In Proceedings of the Fourteenth International Conference on Tangible, Embedded, and Embodied Interaction (TEI '20), Sydney, Australia, 9–12 February 2020; pp. 911–915. [CrossRef]
28. Ernst, M.; Girouard, A. Bending Blindly: Exploring Bend Gestures for the Blind. In Proceedings of the 2016 CHI Conference Extended Abstracts on Human Factors in Computing Systems (CHI EA'16), New York, NY, USA, 7–12 May 2016; pp. 2088–2096. [CrossRef]
29. Heo, J.S.; Shishavan, H.H.; Soleymanpour, R.; Kim, J.; Kim, I. Textile-based stretchable and flexible glove sensor for monitoring upper extremity prosthesis functions. *IEEE Sens. J.* **2020**, *20*, 1754–1760. [CrossRef]
30. Michelitsch, G.; Williams, J.; Osen, M.; Jimenez, B.; Rapp, S. Haptic chameleon: A new concept of shape-changing user interface controls with force feedback. In Proceedings of the CHI'04 Extended Abstracts on Human Factors in Computing Systems (CHI EA'04), New York, NY, USA, 24–29 April 2004; pp. 1305–1308. [CrossRef]
31. Burstyn, J.; Banerjee, A.; Vertegaal, R. FlexView: An evaluation of depth navigation on deformable mobile devices. In Proceedings of the 7th International Conference on Tangible, Embedded and Embodied Interaction (TEI'13), New York, NY, USA, 10–13 February 2013; pp. 193–200. [CrossRef]

electronics

MDPI

Article

An Alignment-Free Sensing Module for Noninvasive Radial Artery Blood Pressure Measurement

Binpeng Zhan, Chao Yang, Fuyuan Xie, Liang Hu, Weiting Liu * and Xin Fu

State Key Laboratory of Fluid Power and Mechatronic Systems, School of Mechanical Engineering, Zhejiang University, Hangzhou 310007, China; 11825032@zju.edu.cn (B.Z.); 21825224@zju.edu.cn (C.Y.); 21925162@zju.edu.cn (F.X.); cmeehuli@zju.edu.cn (L.H.); xfu@zju.edu.cn (X.F.)

* Correspondence: liuwt@zju.edu.cn

Abstract: Sensor–artery alignment has always been a significant problem in arterial tonometry devices and prevents their application to wearable continuous blood pressure (BP) monitoring. Traditional solutions are to use a complex servo system to search for the best measurement position or to use an inefficient pressure sensor array. In this study, a novel solid–liquid mixture pressure sensing module is proposed. A flexible film with unique liquid-filled structures greatly reduces the pulse measurement error caused by sensor misplacement. The ideal measuring location was defined as −2.5 to 2.5 mm from the center of the module and the pressure variation was within 5.4%, which is available in the real application. Even at a distance of ±4 mm from the module center, the pressure decays by 23.7%, and its dynamic waveform is maintained. In addition, the sensing module is also endowed with the capability of measuring the pulse wave transmit time as a complementary method for BP measuring. The capability of the developed alignment-free sensing module in BP measurement was been validated. Twenty subjects were selected for the BP measurement experiment, which followed IEEE standards. The experimental results showed that the mean error of SBP is −4.26 mmHg with a standard deviation of 7.0 mmHg, and the mean error of DBP is 2.98 mmHg with a standard deviation of 5.07 mmHg. The device is expected to provide a new solution for wearable continuous BP monitoring.

Keywords: alignment-free; noninvasive; wearable; continuous blood pressure measurement; liquid-filled

Citation: Zhan, B.; Yang, C.; Xie, F.; Hu, L.; Liu, W.; Fu, X. An Alignment-Free Sensing Module for Noninvasive Radial Artery Blood Pressure Measurement. *Electronics* **2021**, *10*, 2896. https://doi.org/10.3390/electronics10232896

Academic Editors: João Paulo Morais Ferreira and Tao Liu

Received: 9 October 2021
Accepted: 21 November 2021
Published: 23 November 2021

1. Introduction

Cardiovascular disease is the leading cause of death in the world [1]. It is estimated that about 270 million people suffer from hypertension in China and nearly 1.13 billion people worldwide [2,3]. Hypertension at any age is associated with a cognitive decline in different abilities [4]. However, prehypertension has few noticeable symptoms, and it is difficult to recognize when lacking frequent blood pressure (BP) examinations because there are many considerable influences on human BP, such as circadian rhythm and the environment. Compared to intermittent BP measurement, continuous BP monitoring is of great significance by providing more comprehensive information for the clinical diagnosis and control of hypertension [5].

As a result of the expansion of health care, commercial wearable BP monitoring devices are already available on the market. For example, Bpump Inc. launched a BP measurement watch (WF1610B) in 2017, and Omron Inc. launched 'HeartGuide' in 2019. However, their measurement principle is still based on the traditional oscillometric method, which cannot meet the requirements of continuous BP monitoring. The clinical available continuous BP monitoring devices include Finometer PRO (Finapres Medical Systems BV, Amsterdam, The Netherlands), BP-8800 (Omron, Colin Co., Ltd., Tokyo, Japan), TL-200 (Tensys Medical Inc., CA, USA), and BPro (Healthstats International Pte. Ltd., Singapore). Their servo-mechanism makes them bulky and prevents their application in daily life.

Research on noninvasive wearable continuous BP measurement is ongoing. For measuring methods, photoplethysmogram (PPG), arterial tonometry, and pulse wave transit time (PWTT) methods have been widely studied [6]. Several scholars have contributed to modern BP measurements from theoretical and algorithmic perspectives, such as the establishment of multifactor mathematical models [7–9], the use of the adaptive filter to reduce motion artifacts [10,11], and the application of machine learning to assist in calculations [12–14]. Another research field is the enhancement of the sensing device. Nonetheless, a miniaturized, alignment-free, and high sensitivity sensor has always been desired [15]. However, noninvasive BP monitoring is still difficult to accomplish with current sensors. In addition, the sensors are generally specialized, so it is not easy to combine different measurement methods. Specifically, the PPG method is always matched with a photoelectric sensor, which is energy intensive; sensors for the PWTT method mostly use a distributed placement and are challenging to integrate; and for arterial tonometry, a pressure sensor is commonly used. Some scholars have presented excellent solutions with developing technologies, such as liquid capsule structures [16], ultrasonic patch devices [17], and a smartphone-based oscillometric finger-pressing method [18]. In the areas of noninvasive, portable, and continuous measurement, there is still much space for progress.

In this study, the authors proposed an alignment-free compact sensing module. The main obstacle to arterial tonometry is the rigorous placement of the sensor. Therefore, the sensing module was designed as a solid–liquid composite structure, which aims at reducing the pressure measurement error caused by misplacement. Furthermore, it has the potential to combine the PWTT method with arterial tonometry. In order to demonstrate its BP monitoring capability in daily life, the sensing module was integrated into a wearable BP device. The performance of the device was systematically verified by simulation and experiment.

This paper is organized as follows. The theoretical method and device development process are presented in Section 2. The experimental results are described and discussed in Section 3. Finally, the conclusion and prospects are summarized in Section 4.

2. Device Development

The radial artery BP measurement device includes a wrist wearable device and a smartphone-based arterial blood pressure (ABP) recording APP, as shown in Figure 1. Data transmission between the wearable device and the recording APP is via Bluetooth Low Energy (BLE), which removes the need for an enormous data acquisition system, and satisfying the need for the portability of continuous BP measurement in daily life. The measuring principle of the device is mainly based on that of arterial tonometry, combined with the PWTT measurement.

2.1. Theory

Arterial tonometry is a noninvasive BP measurement method proposed by Pressman and Newgard in 1963 [19]. A cross-sectional view of the human wrist is shown in Figure 2a. The radial artery is located between the epidermal tissue and the radius, flanked by the flexor and extensor muscles. After wearing the measurement device, the changes in the radial artery are shown in Figure 2b. Due to the squeezing of the wearable device, the sensor puts external pressure on the radial artery. As a result, the epidermal tissue, radial artery, and radius are close to each other, leading to deformation of the radial artery. When the external pressure is appropriate, the radial artery section will become flat, as shown in Figure 2c. In this case, the tension of the arterial wall is completely horizontal; thus, pressure measured by the sensing module will be equal to the pressure in the artery (BP) if the elastic attenuation of the epidermal tissue is neglected.

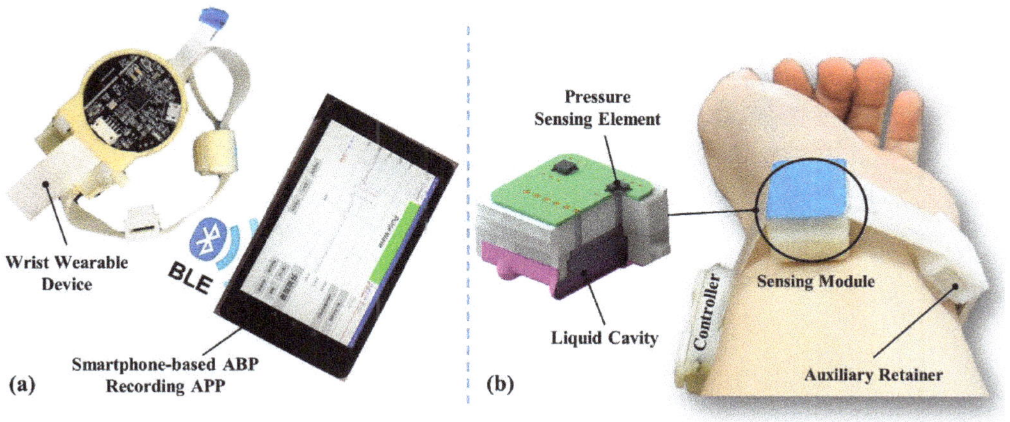

Figure 1. The radial artery BP measurement device: (**a**) BP measurement device and smartphone-based recording APP; (**b**) photo of the device worn on the wrist.

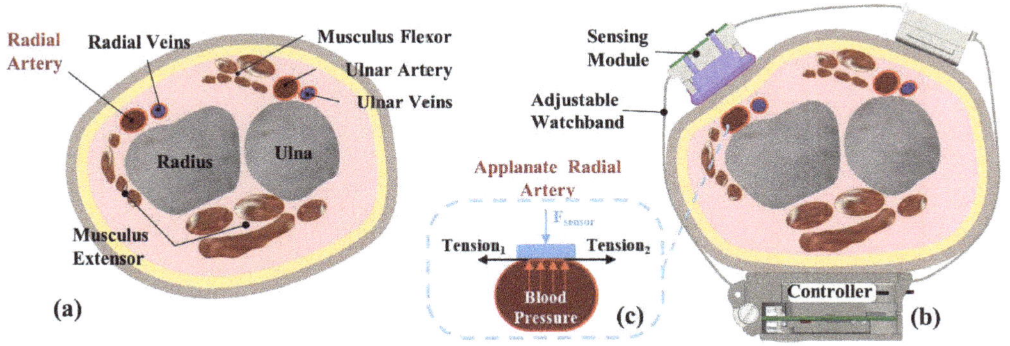

Figure 2. Schematic of arterial tonometry: (**a**) cross section of the human wrist; (**b**) device wearing schematic; (**c**) applanate radial artery.

According to the principle of arterial tonometry, the module has to be directly above the radial artery and remain fixed during measurement. Otherwise, it is possible to introduce considerable measuring errors. However, it is relatively difficult to align the center of the module to the central line of underneath the vascular vessel, whose diameter is relatively small (about 3 mm). Usually, the sensor alignment can only be roughly determined by feeling the pulse with the finger.

2.2. Sensing Module Design

The size of the sensing module is 16 mm × 14 mm, and the width of the module is considered as the distance between the radial protrusion and the musculus flexor, because it is convenient to find the appropriate circumferential position of the module with the help of the radial protrusion and musculus flexor. However, the distance between the radial protrusion and the flexor muscle is an individualized parameter. In order to cover the different ages and weights of all possible users, so that they can easily align the sensing module to the artery, the module should still be positionally robust. The wrist circumference in adults is typically between 140 and 180 mm, and the distance difference between the radial protrusion and the musculus flexor is less than 8 mm. Therefore, the module needs to be robust with a minimum of 8 mm tolerance in the width direction, i.e., the signal measured within this range should be distortion-free and genuinely reflect the pulse state if motion artifacts are ignored.

Inspired by the principle of finger-feeling pulse in traditional Chinese medicine [20], we designed a solid–liquid composite sensor structure, and details are shown in Figure 3. Inside the sensing module is a cavity as shown in Figure 3b, which is filled with silicone oil and connected to the outside through two channels above, one for liquid injection and the other for pressure transfer. The sensitive areas are the two semicylindrical bulges on the bottom of the module as shown in Figure 3c. When a stimulus is received, the bulges deform and generate compressive stress, which can be efficiently transmitted to the pressure sensing element according to Pascal's law. Theoretically, regardless of where the bulges are pressed, the pressure transmitted to the sensing element should be the same.

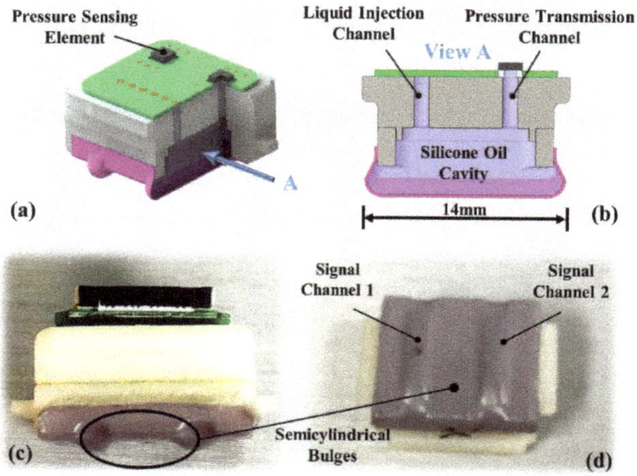

Figure 3. Sensing module details: (**a**) module structure schematic; (**b**) cross-section of the cavity; (**c**) side view of the sensing module; (**d**) semicylindrical bulge of the two channels.

The designed sensing module is a dual-channel signals sensor, as shown in Figure 3d, which can measure the pulse at two adjacent locations. The first purpose is to facilitate the alignment of the module in the axial direction of the wrist because at least one of the two channels can be supported by the radius. The second is that the PWTT can be derived by the time difference between the pulse signal in two different locations. PWTT is a BP-related variable and can be used to derive BP [21]. The result calculated by PWTT can be cross-validated with arterial tonometry.

2.3. Fabrication

The sensitive area of the sensing module, i.e., the area in contact with the skin of the wrist, is made of a thermoplastic polyurethane (TPU) film with good plasticity and biocompatibility. The cavity is filled with non-volatile and biocompatible dimethyl silicone oil (Dow Corning Inc., Michigan, USA, pmx-200). The pressure sensing element of the module is a silicon piezoresistive die (Uni Sense Technology Co. Ltd., Shenzhen, China, US9173) based on micromachining technology.

The preparation process of the sensing module is shown in Figure 4. First, the TPU film with two semicylindrical bulges was produced by hot extrusion forming through a mold. Then the formed film was glued together with the 3D-printed base. Second, the pressure sensing element was glued to the base in alignment with the pressure transmission channel. Third, the cavity was filled with degassed silicone oil through the liquid injection channel in a vacuum environment, and then the liquid injection channel was sealed. Finally, after cleaning the pressure sensing element, it was connected to the peripheral circuit using ultrasonic gold wire bonding technology. The sensing module prepared by this process overcomes the disadvantages of stress transfer attenuation and non-uniform stress distribution in the traditional elastic layer.

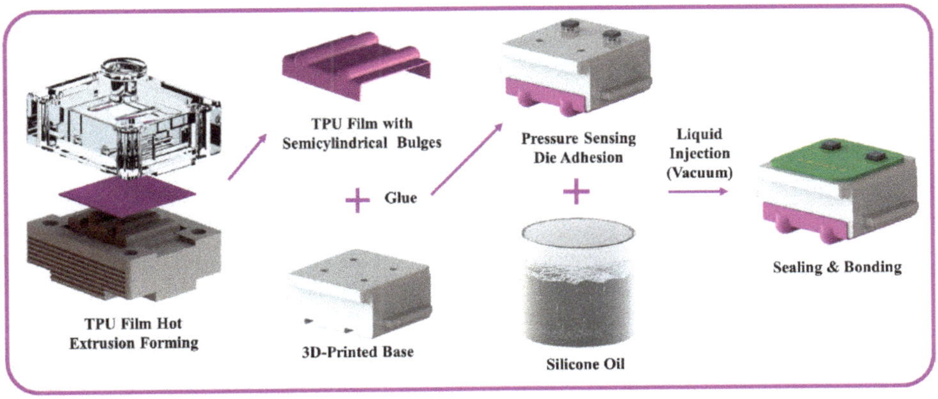

Figure 4. Preparation process of the sensing module.

2.4. Signal Process

The processing flow of the signal collected by the sensing module is shown in Figure 5. The lower left of the figure is a photo of the processing circuit, which integrates power management, signal acquisition, signal processing, and BLE communication functions. It is mounted in the controller with the battery underneath as shown in Figure 2b.

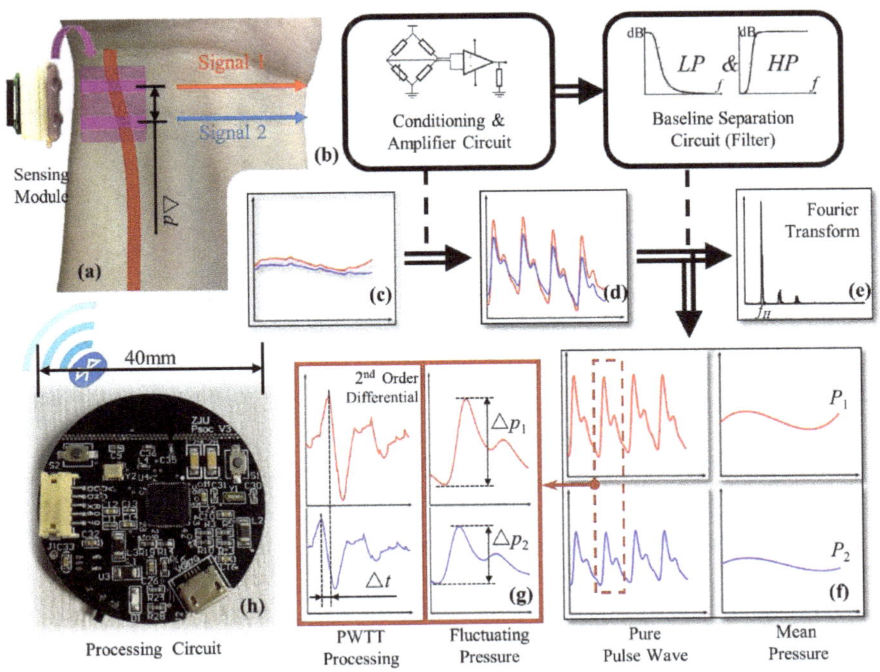

Figure 5. Signal processing flow chart: (**a**) sensing module; (**b**) schematics of conditioning, amplifier, and baseline separation circuit; (**c**) raw pulse signal; (**d**) amplified pulse signal; (**e**) heart rate processing; (**f**) pure pulse wave signal and the mean pressure signal; (**g**) PWTT processing; (**h**) photo of the processing circuit.

The raw signal measured by the sensing module is insignificant, as shown in Figure 5c. First, a distinct pulse signal is obtained through a conditioning and amplifier circuit. Then the heart rate f_H can be obtained by Fourier transform of the pulse signal and taking the dominant frequency, as shown in Figure 5e. The baseline separation circuit consists of a high-pass filter and a low-pass filter circuit. After passing through the two filters separately, two pure pulse wave signals and two mean pressure signals are available; the typical signals are shown in Figure 5f. The mean pressure of the two channels is defined as P_1 and P_2, respectively. The mean pressure here is influenced by the mean BP (MBP) and external pressure, so it can be used to extract the MBP when the external pressure is controllable. The oscillating pressure of the two channels, i.e., the peak-to-peak values of the pure pulse wave signals, are defined as ΔP_1 and ΔP_2, respectively. The oscillating pressure can be used to extract the difference between systolic and diastolic pressure. Then, the 2nd order differential sequence of the pulse wave is derived, and the moment when it reaches its peak is considered as the initial ejection point of the heart [22]. The pulse wave transmit time (PWTT) in the artery can be obtained as the time difference Δt between the initial ejection time of two channels, as shown in Figure 5g. Moreover, the pulse wave velocity (PWV) in the artery can be calculated according to the distance Δd and PWTT between the two signal channels of the sensing module.

Through wireless BLE communication, the measured information can be synchronized to the smartphone for saving and further processing, enabling real-time monitoring of arterial blood pressure.

3. Experiment and Discussion

3.1. Simulation Results on Static Load

In this section we use finite element analysis (FEA) to demonstrate that the structure of liquid sealed in a film allows the pressure to be uniformly transmitted to the pressure sensing element, realizing the insensitivity to the stimulus location. The static simulation analysis was performed based on ANSYS. A simulation model for the basic structure of the sensing module was built first, and the same load was applied to different positions on the bulge structure, as shown by the arrows in Figure 6a. The red arrow is the center loading position, and the yellow arrows are the loading positions when the load is offset to the left by 0.5 mm each time (symmetric on the right). The pressure value at the sensing element was calculated separately for each loading position and defined as the output pressure.

The calculated results of the FEA are shown by the red curve in Figure 6b. The *Y*-axis of the curve is the output pressure attenuation relative to the central loading position (red arrow) when loading at other positions (yellow arrows). Results indicate that the output pressure varies with the position of the load, and the attenuation becomes more pronounced away from the center. The output pressure is attenuated by 5.6% at 2.5 mm away from the center and 25% at 4 mm. This means that the output pressure is nearly the same when the stimulus position is within 5 mm of the middle of the sensing module, and the output pressure error is less than 25% within 8 mm. Even at the furthest edge, the sensing module can obtain a valid signal within 30% attenuation. By consideration of adult vessel size, 2.5 mm away from the center is the location that the whole vessel maintains above the sensing bulge; more misplacement leads to the vessel being only partly above the sensing bulge, which results in a larger measuring error. Fortunately, the approximate pulse finger feeling method can easily guarantee that the module is fixed within the $\pm$2.5 mm misplacement tolerance range.

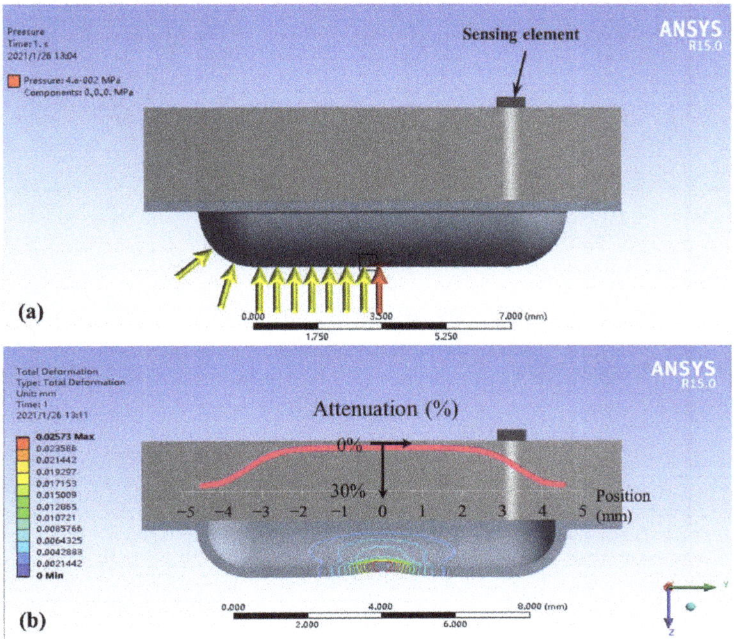

Figure 6. Finite element simulation: (**a**) loading positions; (**b**) results of output pressure and film deformation.

3.2. Location Robustness on Dynamic Stimulation

The location robustness in a static load was verified in the simulation analysis. However, the effects of pulse oscillation and epidermal tissue elasticity on pressure signal were not considered in the simulation. Therefore, in this section, the response to the stimulus location of the sensing module under near-real and controlled dynamic conditions is experimentally investigated. The schematic diagram of the experimental platform is shown in Figure 7a. A peristaltic pump was used to simulate the heartbeat, a rubber hose to simulate the blood vessel, and a silica gel pad to simulate the epidermal tissue. The peristaltic pump rollers squeeze the rubber hose from left to right in sequence, intermittently pumping the water through the region where the sensing module is located. The diameter of the rubber hose was 3 mm, which is similar to that of the radial artery vessel. The thickness of the silica gel pad was 5 mm and the hardness was 5 HA, similar to those of the epidermal tissue. The rigid substrate beneath the silica gel pad was used to simulate the radius supporting the radial artery. Figure 7b shows the right elevation of the sensing module and the rubber hose, which illustrates the change in the rubber hose location relative to the sensing module from −5 to 5 mm during the experiment. The photo of the experimental platform is shown in Figure 7c, including a peristaltic pump, a segment of rubber hose, a silica gel pad, a sensing module, a processing circuit, and a smartphone-based recording APP. Figure 7d–f show the close-up photos of the rubber hose positions at the leftmost (−5 mm), middle (0 mm), and rightmost (5 mm), respectively.

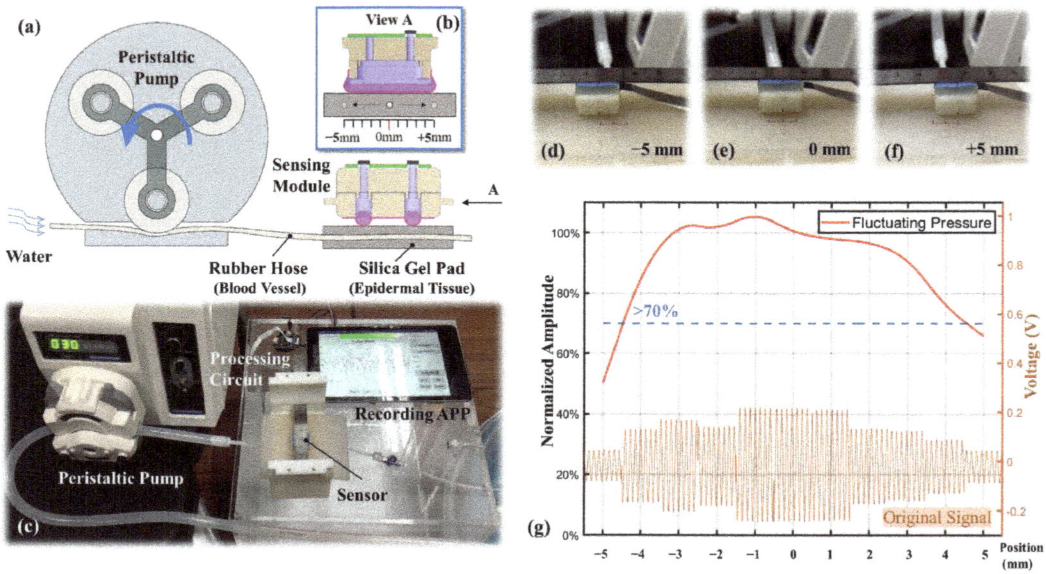

Figure 7. Experimental setup for location robustness: (**a**) schematic diagram of the experimental platform; (**b**) right elevation of the sensing module; (**c**) photo of the experimental setup; (**d–f**) different test locations; (**g**) results of the location robustness experiment.

The rotational speed of the peristaltic pump is set to 30 rpm and with the three rollers in one circle, i.e., the fluctuation frequency is 90 Hz, which is close to the human heart rate. As each roller passes through the rubber hose, the peak pressure is generated. The signals collected at each location of the sensing module are shown at the bottom of Figure 7g. The top of the figure shows the extracted average oscillating pressure versus its location, and the Y-axis is the oscillating pressure attenuation relative to the central location. As can be seen from the figure, the dynamic experimental results are almost the same as the static simulation results. At a distance of ± 4 mm from the center, the oscillating pressure decays by 23.7%. The ideal measuring location is from -2.5 to 2.5 mm and the oscillating pressure variation is within 5.4%. Another important result is that the waveforms maintained a close agreement at all the misalignment locations, which is one of the key performances desired in real BP monitoring. In this manner, the designed sensing module demonstrated its capability of alignment-free tonometry BP measurement.

Based on the developed alignment-free sensing module, it was necessary to examine its appropriate capability in BP measuring, which is presented in the following sections.

3.3. Individualized Calibration

In this section, the arterial blood pressure (ABP) measured by the sensing module is investigated. The subject's ABP was artificially and quantitatively altered through an in vivo BP control platform to establish a link with the output of the sensing module. This calibration method combines ABP with hydrostatic pressure and is able to assess the effect of individualization factors, such as vessel thickness and epidermal tissue stiffness. The calibration results can be applied to BP tracking for long-term wear. In this study, artificial alteration of ABP was achieved by adjusting the height of the subject's wrist relative to the heart. According to Bernoulli's principle, for every 1.3 cm increase in the height of the wrist relative to the heart, the ABP at the measuring point decreases by 1 mmHg. However, due to the difference in physiological characteristics of the individual body, the signal collected by the sensing module varies from person to person even if the

ABP variation is the same. Therefore, it is necessary to recalibrate the sensitivity by the controlled ABP and the corresponding sensor output.

The experimental setup includes a wrist height adjustment platform, as shown in Figure 8a, which is designed with a sliding track allowing the subjects to maintain the posture while changing the height of the wrist. The subject selected for the experiment was healthy, did not smoke, drink, or exercise before the experiment, and was in a resting state without any stimulus during the experiment. Therefore, the BP of the subject was considered to be stable and did not fluctuate significantly during the experiment. The experimental procedure started with finding the height of the heart and defining it as 0 cm. The wrist height was adjusted within ±20 cm, starting from −20 cm and rising 10 cm each time. Each position measurement lasted 30 s, and the measured pulse wave is shown in Figure 8b.

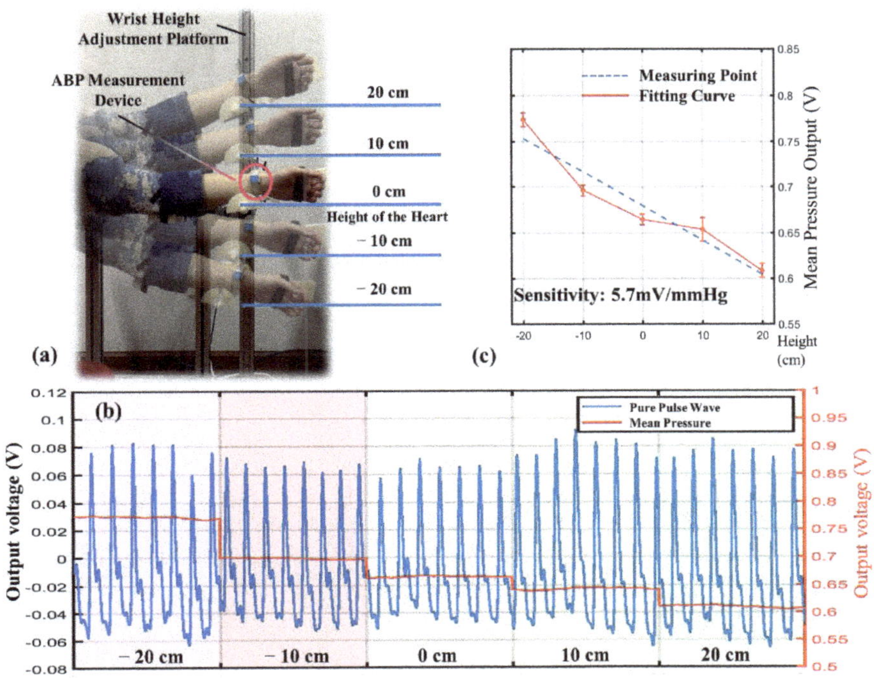

Figure 8. Individualized calibration results. (**a**) Experimental setup for ABP control. (**b**) Pulse signals at different heights. (**c**) Recalibrated sensitivity.

By the analysis of the experimental data, it can be found that the primary change in the signal as the wrist height rises is the decrease in mean pressure. The change rate is −4.25 mV/cm; that is, the sensitivity of the sensing module is 5.7 mV/mmHg. However, the sensitivity of the sensor was previously calibrated as 8.7 mV/mmHg in a precise pressure chamber. The error between the measured and calibrated sensitivity is a reflection of the individualization factor. This can be explained in two ways. One is that the epidermal tissue induces an attenuation effect on pressure transmission [23,24]. The other is that only part of the sensitive area is in contact with the arterial blood vessels during the measurement, as shown in Figure 5a. The first factor is related to the elasticity and thickness of the epidermis, and the second factor is influenced by the thickness of the artery. Both factors can be attributed to the individualized discrepancies and can be eliminated by recalibration.

3.4. BP Calibration

The individualized factor calibration characterizes the relationship between the variation in the sensing module output and BP during ambulatory BP measurements. However, because the wearing state is different each time, the initial BP also needs to be calibrated. The characteristics of the transmural pressure can be used to calibrate the initial BP. The transmural pressure represents the combined pressure of intravascular pressure, applied pressure, and hydrostatic pressure. It is generally accepted that the compliance of the vessel increases to a maximum value as the transmural pressure goes to zero [25]. In addition, the transmural pressure equal to zero means that the intravascular pressure, i.e., the MBP, is equal to the extravascular pressure when the hydrostatic pressure is equal to zero. This also means that the pulse oscillating pressure measured by the sensing module is the largest [26]. Furthermore, maximum vascular compliance also means that the same pressure change in the vessel causes the most significant change in vessel volume and, therefore, the most extended PWTT [27]. Thus, there can be two ways to determine the initial BP—when the pulse oscillation amplitude reaches its maximum and when the PWTT is the longest. In this study, the transmural pressure was varied by changing the applied pressure of the sensing module, and was decreased to zero and then increased in reverse as the applied pressure increases.

The experimental setup containing a pressure controller, a gasbag, and an ABP measurement device is shown in Figure 9a. The pressure applied to the wrist by the sensing module was controlled by the pressure controller and a gasbag above the module. Before the experiment, the subject was required to align the sensing module with the gasbag after wearing the ABP measurement device, and the subject's wrist remained stationary throughout the experiment. The pressure on the wrist was then controlled to increase gradually and uniformly until the acquired pulse waveform begins to distort, indicating that the artery is almost closed.

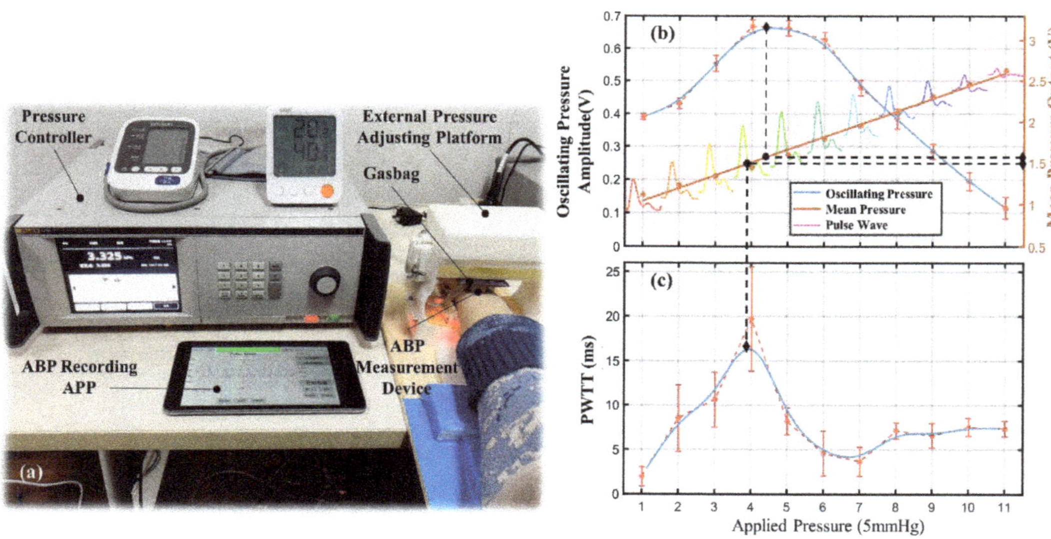

Figure 9. Pulse characteristics at different applied pressures.(**a**) Experimental setup for initial BP calibration. (**b**) Mean pressures and oscillating pressures. (**c**) PWTT.

The experimental results for one subject are presented in Figure 9b. The mean pressure measured by the sensing module keeps rising uniformly with the increase in the applied pressure. The oscillating pressure and the PWTT also progressively increase at the beginning due to the decreased transmural pressure. As the applied pressure continued to increase, the oscillating pressure and the PWTT started to decrease from a certain point because the transmural pressure decreased to zero and then increased in reverse. The turning point is the MBP state for the subject, as shown by the dashed line in Figure 8c. The mean pressure at this point corresponds to the MBP, and the oscillating pressure corresponds to the difference between systolic and diastolic pressures.

According to the IEEE standard [28], twenty healthy adults were selected for evaluating the effectiveness of the designed BP measurement device. Their personal information and BP measured by a commercial sphygmomanometer (Omorn Inc., Japan, HEM-7124) 5 min before the test are shown in Table 1. The initial BP was calibrated by the same method for the 20 subjects, and the extracted mean and oscillating pressures are presented in Figure 10. The calibration results of the MBP are shown in red.

Because the BP procedure in this study is based on arterial tonometry, the MBP was obtained from the corresponding mean pressure, and the differential pressure (DP) between systolic pressure and diastolic pressure was obtained from the corresponding oscillating pressure. The BP measurement results of the subject were determined by the MBP and the DP. In order to enable comparison with the commercial cuff sphygmomanometers, the results were converted to SBP and DBP by the 4/6 principle. Then, the Bland–Altman method was used to analyze the correlation between the derived BP and the BP measured by the commercial sphygmomanometer; the results are shown in Figure 11. All measuring results fall within the confidence interval. The mean error of SBP is −4.62 mmHg with a standard deviation of ±7.0 mmHg, and the mean error of DBP is 2.98 mmHg with a standard deviation of ±5.07 mmHg. The BP measurement results are in accordance with the AAMI standard [29] of 5 ± 8 mmHg. The above experimental results indicate that the BP measurement device proposed in this paper has considerable stability and can adapt to different people.

Table 1. BP information for the 20 subjects.

Subject	SBP (mmHg)	DBP (mmHg)	Gender	Age	Weight (kg)
1	120	73	Male	23	64
2	116	75	Male	22	57
3	115	61	Male	22	65
4	125	85	Male	21	76
5	116	74	Male	22	75
6	113	84	Male	22	72
7	122	76	Male	22	50
8	111	69	Female	22	62
9	115	78	Male	24	60
10	102	75	Female	22	46
11	122	74	Male	22	100
12	129	85	Male	23	61
13	137	84	Male	22	70
14	106	76	Male	23	70
15	128	83	Male	23	70
16	99	69	Male	22	65
17	121	78	Male	23	79
18	104	79	Male	23	78
19	101	71	Male	23	69
20	97	64	Male	23	70

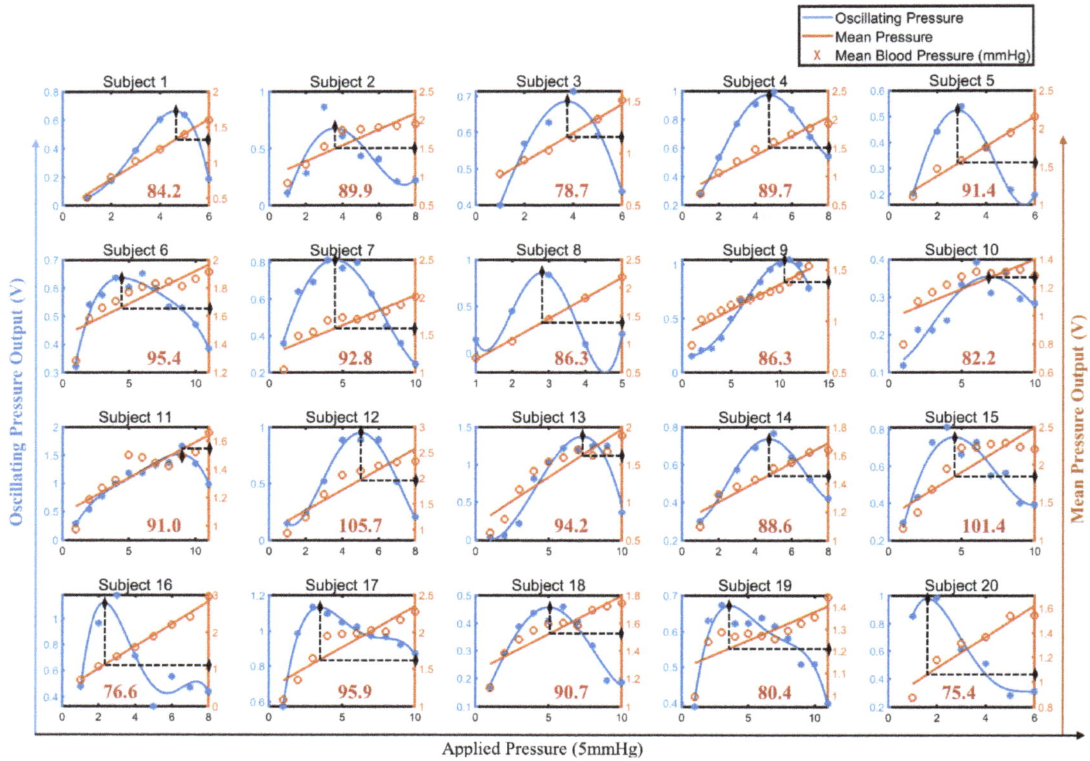

Figure 10. Initial BP caalibration results for the 20 subjects.

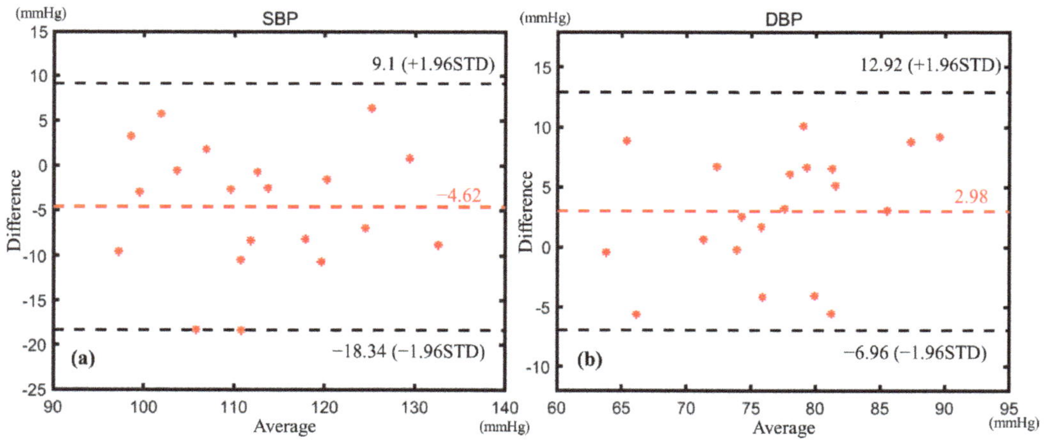

Figure 11. Bland–Altman analysis of (**a**) SBP and (**b**) DBP.

4. Conclusions

Arterial tonometry is a noninvasive BP measurement method with high accuracy. However, sensor–artery alignment is a major problem that obstructs the application of arterial tonometry. In this study, a novel solid–liquid mixture pulse sensing module was proposed to address the existing problems. The flexible film with semicylindrical bulges and unique liquid-filled structure greatly reduces the pulse measuring error caused by position deviation. Having a rational circuit and algorithm design, it is able to extract the mean and oscillating pressures of the subject's pulse. In addition, the device has the ability to measure PWTT, which can serve as a complement to arterial tonometry and is especially suitable for ambulatory BP monitoring. The location robustness of the sensing module was verified by simulation and experiment. The ideal measuring location ranges from −2.5 to 2.5 mm of the module and the pressure variation is within 5.4%, which can be easily achieved by finger feeling in a real application of a wearable BP monitoring scenario. At a distance of ±4 mm from the module center, although the pressure decays by 23.7%, the dynamic waveform is conserved well, which is important for wearable application. For different potential users, the individualization factor can be calibrated by an ABP control platform, and the initial BP can be calibrated by an applied pressure regulation platform. The alignment deviation errors can be further eliminated by the individual calibration procedure in a practical BP measuring step. BP measurement experiments were performed on 20 subjects, and the experimental procedure followed IEEE standards. The results indicate that the wearable device performs well for BP measurements and the error of the results meet the AAMI standards. The device is expected to provide a new solution for wearable continuous BP monitoring.

In next phase, we plan to (1) combine arterial tonometry and the PWTT method for BP measurement, and use the device for noninvasive continuous BP measurement in clinical trials; (2) study the interference of human motion on the measurement, and improve the accuracy of the measurement in motion in terms of the device structure design and an anti-motion algorithm.

Author Contributions: Conceptualization, W.L.; software, B.Z.; validation, B.Z., C.Y. and F.X.; formal analysis, B.Z.; data curation, C.Y. and F.X.; writing—original draft preparation, B.Z.; writing—review and editing, W.L.; supervision, W.L. and X.F.; funding acquisition, W.L., L.H. and X.F. All authors have read and agreed to the published version of the manuscript.

Funding: This research work was financially supported by the National Natural Science Foundation of China, grant number 51875506, the Science Fund for Creative Research Groups of National Natural Science Foundation of China, grant number 51821093, the National Natural Science Foundation of China, grant number 51775485.

Institutional Review Board Statement: The study was conducted according to the guidelines of the Declaration of Helsinki, and approved by the Medical Ethics Committee of College of Biomedical Engineering & Instrument Science, Zhejiang University ([2021]-39).

Informed Consent Statement: Informed consent was obtained from all subjects involved in the study.

Acknowledgments: The authors are also sincerely grateful to the students participating in the experiment.

Conflicts of Interest: The authors declare no conflict of interest.

References

1. Kannel, W.B.; Mcgee, D.L. Kannel WB and McGee DL. Diabetes and cardiovascular disease: The Framingham study. *J. Am. Med. Assoc.* **1979**, *241*, 2035–2038. [CrossRef]
2. Ma, L.Y.; Chen, W.W.; Gao, R.L.; Liu, L.S.; Zhu, M.L.; Wang, Y.J.; Wu, Z.S.; Li, H.J.; Gu, D.F.; Yang, Y.J.; et al. China cardiovascular diseases report 2018: An updated summary. *J. Geriatr. Cardiol.* **2020**, *17*, 1–8. [CrossRef]
3. Roth, G.A.; Mensah, G.A.; Johnson, C.O.; Addolorato, G.; Ammirati, E.; Baddour, L.M.; Barengo, N.C.; Beaton, A.Z.; Benjamin, E.J.; Benziger, C.P.; et al. Global Burden of Cardiovascular Diseases and Risk Factors, 1990–2019. *J. Am. Coll. Cardiol.* **2020**, *25*, 2982–3021. [CrossRef]
4. de Menezes, S.T.; Giatti, L.; Brant, L.C.; Griep, R.H.; Schmidt, M.I.; Duncan, B.B.; Suemoto, C.K.; Ribeiro, A.L.; Barreto, S.M. Hypertension, Prehypertension, and Hypertension Control: Association With De-cline in Cognitive Performance in the ELSA-Brasil Cohort. *Hypertension* **2020**, *77*, 80. [CrossRef]
5. Turner, J.R.; Viera, A.J.; Shimbo, D. Ambulatory Blood Pressure Monitoring in Clinical Practice: A Review. *Am. J. Med.* **2014**, *128*, 14–20. [CrossRef]
6. Rastegar, S.; GholamHosseini, H.; Lowe, A. Non-invasive continuous blood pressure monitoring systems: Current and proposed technology issues and challenges. *Phys. Eng. Sci. Med.* **2019**, *43*, 11–28. [CrossRef]

7. Singh, P.; Choudhury, M.I.; Roy, S.; Prasad, A. Computational Study to Investigate Effect of Tonometer Geometry and Pa-tient-Specific Variability on Radial Artery Tonometry. *J. Biomech.* **2017**, *58*, 105–113. [CrossRef]
8. Kachuee, M.; Kiani, M.M.; Mohammadzade, H.; Shabany, M. Cuffless Blood Pressure Estimation Algorithms for Continuous Health-Care Monitoring. *IEEE Trans. Biomed. Eng.* **2016**, *64*, 859–869. [CrossRef]
9. Poon, C.C.Y.; Zhang, Y.-T.; Liu, Y. Modeling of Pulse Transit Time under the Effects of Hydrostatic Pressure for Cuffless Blood Pressure Measurements. In Proceedings of the 2006 3rd IEEE/EMBS International Summer School on Medical Devices and Biosensors, Cambridge, MA, USA, 4–6 September 2006; Volume 36, pp. 65–68. [CrossRef]
10. Ram, M.R.; Madhav, K.V.; Krishna, E.H.; Komalla, N.R.; Reddy, K.A. On the performance of AS-LMS based adaptive filter for reduction of mo-tion artifacts from PPG signals. In Proceedings of the 2011 IEEE International Instrumentation and Measurement Technology Conference, Hangzhou, China, 10–12 May 2011; pp. 1–4.
11. Ram, M.R.; Madhav, K.V.; Krishna, E.H.; Komalla, A.R.; Reddy, K.A. A Novel Approach for Motion Artifact Reduction in PPG Signals Based on AS-LMS Adaptive Filter. *IEEE Trans. Instrum. Meas.* **2011**, *61*, 1445–1457. [CrossRef]
12. Kurylyak, Y.; Lamonaca, F.; Grimaldi, D. A Neural Network-based method for continuous blood pressure estimation from a PPG signal. In Proceedings of the 2013 IEEE International Instrumentation and Measurement Technology Conference (I2MTC), Minneapolis, MN, USA, 6–9 May 2013; pp. 280–283.
13. He, R.; Huang, Z.P.; Ji, L.Y.; Wu, J.K.; Li, H.; Zhang, Z.Q. Beat-to-beat ambulatory blood pressure estimation based on random forest. In Proceedings of the IEEE International Conference on Wearable & Implantable Body Sensor Networks, Chicago, IL, USA, 19–22 May 2019; pp. 194–198.
14. Miao, F.; Fu, N.; Zhang, Y.; Ding, X.-R.; Hong, X.; He, Q.; Li, Y. A Novel Continuous Blood Pressure Estimation Approach Based on Data Mining Techniques. *IEEE J. Biomed. Health Inform.* **2017**, *21*, 1730–1740. [CrossRef]
15. Heikenfeld, J.; Jajack, A.; Rogers, J.; Gutruf, P.; Tian, L.; Pan, T.; Li, R.; Khine, M.; Kim, J.; Wang, K. Wearable sensors: Modalities, challenges, and prospects. *Lab. Chip.* **2018**, *18*, 217–248. [CrossRef]
16. Fan, X.; Huang, Y.; Ding, X.; Luo, N.; Li, C.; Zhao, N.; Chen, S.-C. Alignment-Free Liquid-Capsule Pressure Sensor for Cardiovascular Monitoring. *Adv. Funct. Mater.* **2018**, *28*, 45. [CrossRef]
17. Wang, C.; Li, X.; Hu, H.; Zhang, L.; Huang, Z.; Lin, M.; Zhang, Z.; Yin, Z.; Huang, B.; Gong, H.; et al. Monitoring of the central blood pressure waveform via a conformal ultrasonic device. *Nat. Biomed. Eng.* **2018**, *2*, 687–695. [CrossRef]
18. Chandrasekhar, A.; Kim, C.S.; Naji, M.; Natarajan, K.; Hahn, J.O.; Mukkamala, R. Smartphone-based blood pressure monitoring via the oscillometric fin-ger-pressing method. *Sci. Transl. Med.* **2018**, *10*, eaap8674. [CrossRef]
19. Pressman, G.L.; Newgard, P.M. A Transducer for the Continuous External Measurement of Arterial Blood Pressure. *IRE Trans. Bio-Med. Electron.* **1963**, *10*, 73–81. [CrossRef]
20. Ni, M. *The Yellow Emperor's Classic of Medicine: A New Translation of the Neijing Suwen With Commentary*; Shambhala: London, UK, 1995.
21. Buxi, D.; Redoute, J.-M.; Yuce, M. A survey on signals and systems in ambulatory blood pressure monitoring using pulse transit time. *Physiol. Meas.* **2015**, *36*, R1–R26. [CrossRef]
22. Hsu, Y.; Young, D.J. Skin-Coupled Personal Wearable Ambulatory Pulse Wave Velocity Monitoring System Using Micro-electromechanical Sensors. *Sens. J. IEEE* **2014**, *14*, 3490–3497. [CrossRef]
23. Liu, W.; Gu, C.; Zeng, R.; Yu, P.; Fu, X. A Novel Inverse Solution of Contact Force Based on a Sparse Tactile Sensor Array. *Sensors* **2018**, *18*, 351. [CrossRef]
24. Woo, S.H.; Choi, Y.Y.; Kim, D.J.; Bien, F.; Kim, J.J. Tissue-Informative Mechanism for Wearable Non-invasive Continuous Blood Pressure Monitoring. *Sci. Rep.* **2014**, *4*, 6618. [CrossRef]
25. Shaltis, P.; Reisner, A.; Asada, H. A hydrostatic pressure approach to cuffless blood pressure monitoring. In Proceedings of the 26th Annual International Conference of the IEEE Engineering in Medicine and Biology Society, Berlin, Germany, 1 September 2004. [CrossRef]
26. Soh, K.-S.; Lee, M.-H.; Yoon, Y.-Z. Pulse type classification by varying contact pressure. *IEEE Eng. Med. Biol. Mag.* **2000**, *19*, 106–110. [CrossRef]
27. Zheng, D.; Allen, J.; Murray, A. Effect of external cuff pressure on arterial compliance. *Comput. Cardiol.* **2005**, *2005*, 315–318. [CrossRef]
28. IEEE Standard Association. *IEEE Standard for Wearable Cuffless Blood Pressure Measuring Devices*; IEEE: Manhattan, NY, USA, 2014; pp. 1–32.
29. Stergiou, G.S.; Palatini, P.; Asmar, R.; Ioannidis, J.P.; Kollias, A.; Lacy, P.; McManus, R.J.; Myers, M.G.; Parati, G.; Shennan, A.; et al. Recommendations and Practical Guidance for performing and reporting validation studies according to the Universal Standard for the validation of blood pressure measuring devices by the Association for the Advancement of Medical Instrumentation/European Society of Hypertension/International Organization for Standardization (AAMI/ESH/ISO). *J. Hypertens.* **2019**, *37*, 459–466. [CrossRef]

electronics

MDPI

Article

A Data Augmentation Method for War Trauma Using the War Trauma Severity Score and Deep Neural Networks

Jibin Yin [1], Pengfei Zhao [1], Yi Zhang [2,*], Yi Han [3] and Shuoyu Wang [3]

[1] Faculty of Information Engineering and Automation, Kunming University of Science and Technology, Kunming 650500, China; yinjibin@kust.edu.cn (J.Y.); zhaopengfei@stu.kust.edu.cn (P.Z.)
[2] Naval Training Base of Health Service, Naval Medical University, Shanghai 200433, China
[3] School of System Engineering, Kochi University of Technology, Kami City 780-8520, Japan; 258012g@gs.kochi-tech.ac.jp (Y.H.); wang.shuoyu@kochi-tech.ac.jp (S.W.)
* Correspondence: 18602160005@126.com

Abstract: The demand for large-scale analysis and research of data on trauma from modern warfare is increasing day by day, but the amount of existing data is not sufficient to meet such demand. In this study, an integrated modeling approach incorporating a war trauma severity scoring algorithm (WTSS) and deep neural networks (DNN) is proposed. First, the proposed WTSS, which uses multiple non-linear regression based on the characteristics of war trauma data and the medical evaluation by an expert panel, performed a standardized assessment of an injury and predicts its trauma consequences. Second, to generate virtual injury, based on the probability of occurrence, the injured parts, injury types, and complications were randomly sampled and combined, and then WTSS was used to assess the consequences of the virtual injury. Third, to evaluate the accuracy of the predicted injury consequences, we built a DNN classifier and then trained it with the generated data and tested it with real data. Finally, we used the Delphi method to filter out unreasonable injuries and improve data rationality. The experimental results verified that the proposed approach surpassed the traditional artificial generation methods, achieved a prediction accuracy of 84.43%, and realized large-scale and credible war trauma data augmentation.

Keywords: artificial intelligence; data augmentation; war trauma severity score; deep neural network

Citation: Yin, J.; Zhao, P.; Zhang, Y.; Han, Y.; Wang, S. A Data Augmentation Method for War Trauma Using the War Trauma Severity Score and Deep Neural Networks. *Electronics* **2021**, *10*, 2657. https://doi.org/10.3390/electronics10212657

Academic Editor: Enzo Pasquale Scilingo

Received: 16 September 2021
Accepted: 26 October 2021
Published: 29 October 2021

1. Introduction

War trauma data are the core elements of wargaming, military medical service training, and medical decision-making [1]. With the continuous development of modern warfare, the analysis and research of physical war trauma data have become more and more important. However, the amount of existing data is not sufficient to support large-scale analysis and evaluation, and the confidential nature of war trauma data makes them hard to collect and obtain from public channels. Therefore, efficient and credible data augmentation of war trauma data has become a research work with great practical significance. To the best of our knowledge, research on this topic has been limited. In the currently used method, the additional physical trauma data are still artificially generated by well-trained experts or doctors based on their professional knowledge and experience. However, this method is not only inefficient, time-consuming, and labor cost-intensive, but also inherently biased due to its dependence on personal subjective cognition, which is difficult to overcome. In addition, different experts have no unified standard for assessing injury consequences. Furthermore, the amount of artificially generated war trauma data is too small to meet the actual needs. Therefore, we developed a standardized evaluation algorithm to improve the quality of assessment of injury consequences and find an automatic, efficient, and credible approach for small-sample augmentation of war trauma data.

More than half a century since the concept of artificial intelligence (AI) was first formally proposed at the Dartmouth Conference [2], the AI technology has empowered amazing developments in many fields. Meanwhile, the external environment and challenges faced by the development of AI have also undergone profound changes [3]. These changes are especially prominent in certain fields, such as big data, virtual reality, supercomputing, and mobile payment. Therefore, under the trend that the overall environment is getting closer to big data, deep learning (DL), which is based on machine learning, has become the core element of the application of AI [4] and has led to satisfactory application results in many fields, such as cloud computing [5], image identification [6], sports training [7], and AlphaGo [8]. Recently, AI technologies such as DL started to be gradually applied in the field of medical research, including in promoting disease management [9], computer-aided diagnosis [10], biomedical information processing [11], medical image recognition [12], and disease prediction [13]. Especially in disease prediction, AI has been recognized as one of the key elements of an accurate and robust prediction system [14]. For example, deep neural networks (DNNS), which are AI tools, are now used to assist physicians and for automatic diagnosis. Specific application cases include early detection of cardiovascular disease [15], cancer diagnosis [16], survival prediction [17], and injury severity assessment [18].

Compared with machine-learning methodologies and shallow neural networks, DL, which is now the core of the AI method, overcomes the research drawbacks of limited samples and low generalizability by training large-scale annotated sample data to automatically extract complex sample features and fully optimize the model parameters layer by layer. Thus, DL can carry out a more essential characterization of the data and demonstrates a superior feature-learning ability [19]. In other words, with the existing technology level, the larger the scale and the higher the quality of the annotated data are, the better the performance of the model will be. Therefore, DL can effectively solve many complex problems in the medical field [20,21]. In the prediction and diagnosis of some diseases, the accuracy and efficiency of predictive DL models have surpassed those of professional doctors and experts [22] and have thus made outstanding contributions to the development of the medical field.

2. Related Work

Currently, there are two main methods of data augmentation: oversampling and generative adversarial network (GAN). The principle of oversampling is as follows: if the samples of different classes are imbalanced, the training data can be expanded by copying the training samples of the minority class or adding noises to create new ones [23]. To solve the imbalanced dataset learning problem, in 2002, Bowyer et al. [24] created a synthetic minority oversampling technique (SMOTE), which generated synthetic minority class samples. In 2005, Han et al. [25] proposed a borderline SMOTE algorithm, which considered the minority instances near the borderline and the neighboring instances. The following year, David et al. [26] proposed a cluster SMOTE; Bai et al. [27] proposed an adaptive synthetic sampling approach (ADASYN) for imbalanced learning in 2008; Barua et al. [28] suggested a MWMOTE in 2014; Douzas et al. [29] proposed a SOMO method in 2017. Most of these methods focused on imbalanced learning by adding oversampling examples to the imbalanced datasets. However, physical war trauma data are not imbalanced but insufficient in every class. Therefore, the abovementioned oversampling techniques are not suitable for the augmentation of physical war trauma data.

A GAN is a data augmentation model based on DL, which can be used to learn the potential distribution of complex data, generate large-scale and high-quality synthetic samples, effectively solve the problem of insufficient data due to factors such as difficulty and cost of sample acquisition [30]. Thus, the GAN has become one of the most promising data augmentation approaches in recent years. A GAN is intrinsically a generation model [31] that does not depend on a priori hypotheses but on the internal confrontation between the data and the model itself to achieve unsupervised learning. To solve the inadequate problem of real data, a GAN can generate synthetic samples of the existing data with the same distribution [32]. A GAN's structure consists of two feedforward neural networks:

a generator G and a discriminator D. In the learning process, G continuously generates new synthetic samples while D discriminates between the synthetic samples and the real samples as accurately as possible, then gives feedback. In this way, the GAN has created a game similar to "counterfeit currency identification" in which both sides of the game continue to improve their abilities through confrontation.

However, the samples processed by a GAN are mainly two-dimensional data such as pictures and voice signals. A GAN generates virtual images by rotating, scaling, cropping, and changing the brightness, contrast, hue, saturation and adding random noise to image data. However, the GAN is not a good choice for augmenting physical war trauma data.

In the medical field, the application of medical scoring is increasingly maturing, especially in medical treatment, early diagnosis, trauma assessment, and other aspects to the point that it now plays an important auxiliary role. For example, Gabriele Canzi et al. introduced the comprehensive facial injury (CFI) score for comprehensively evaluating severity of facial injuries [33]. Hasanka Ratnayake et al. used a laboratory-derived early warning score to predict in-hospital mortality and admission to the intensive care unit (ICU) [34]. Konlawij Trongtrakul et al. created the acute kidney injury (AKI) risk prediction score for early prediction of the condition among critically ill surgical patients [35].

The trauma score is a common type of medical score that predicts severity of an injury. It uses scientific scoring to quantitatively or semi-quantitatively assess injury severity and its consequences to the injured [36]. The scoring standard was developed by a panel of experts in the field who will continue to improve and optimize it based on feedback from the application of the trauma score as well as from related research progress. Recently, several improved injury severity score (ISS) methods have been proposed. Cristiane et al. created a novel trauma and injury severity score (TRISS) for survival prediction [37]; Yang et al. used a revised injury severity score (RISS) to evaluate the severity of injuries of patients hospitalized due to an accident [38]; Shi et al. developed a weighted injury severity score (WISS) to improve adult trauma mortality prediction [39]. For example, RISS divides the human body into six public parts: the head, the face, the chest, the abdomen, the limbs, and the body surface. Then, it squares the standard ISS for each of the most serious injuries of the three most serious body parts of the patient and puts them together. As for the second most serious injuries, only their ISS values are put together. If there are more than four injured parts, the standard injury severity score of the most serious injuries of the fourth part is added. The RISS equation is as follows:

$$RISS = (A_1^2 + A_2) + (B_1^2 + B_2) + (C_1^2 + C_2) + D \tag{1}$$

where A_1, B_1, and C_1 mean the most serious injuries of the three most serious body parts; A_2, B_2, and C_2 mean the second most serious injuries of the three most serious body parts; D means the standard injury severity score of the most serious injuries of the fourth part.

Taken together, various novel scientific scoring methods have gradually become doctors' helping hands in evaluating patients' injuries. Medical scoring belongs to the category of predictive science. Because different scoring mechanisms have different limitations, it is impossible to achieve 100% accuracy in prediction. However, with the continuous advancements in medicine and with the revision, expansion, and improvement of the scoring mechanisms by researchers in the related domains, medical scoring approaches are expected to become more scientific, practical, and in line with objective reality [40].

On the other hand, the DL technology combined with knowledge from different disciplines for interdisciplinary field research is an emerging trend. For example, Yang et al. enhanced PIR-based multiperson localization by combining DL with the domain knowledge [41], and Ding et al. combined the domain knowledge and DL for domain adaptation in machine translation [42]. Therefore, combining DL with the domain knowledge of medical experts according to the characteristics of war trauma data is key to the successful application of DL to the augmentation of war trauma data.

Based on the above research, to solve the data augmentation problems with small-sample war trauma data by studying the GAN's idea and the medical trauma scoring method, this article proposes an approach that combines a WTSS with a DNN [43]. The WTSS–DNN integrated model simulates the generative model in thought, including sample generation and discrimination. The injuries are generated through random sampling and evaluated with WTSS, and then marked with an injury consequence label; this is the sample generation link. The assessment of the prediction accuracy of the DNN classifier is combined with the discrimination of unreasonable injuries by the expert panel; this is the discrimination link. After the accuracy and plausibility of the synthetic samples have been judged, the expert panel provides feedback, based on which, on the one hand, the characteristics of the synthetic samples are further investigated while the necessary optimization and adjustments to the WTSS algorithm are made; and on the other hand, the unreasonable synthetic samples are filtered out to improve data rationality. Eventually, the accuracy and plausibility of the augmented data are expected to stabilize and be optimized to generate credible samples.

This data augmentation approach is the first attempt to combine war trauma assessment in the medical field with DL in the AI field. The WTSS–DNN integrated model can automatically generate large-scale and credible virtual war trauma data, making it possible to carry out related data-based military research, which has great practical significance. In addition, this approach not only helps to solve the war trauma data augmentation problem, but the WTSS algorithm we have proposed also provides a practical auxiliary tool for quickly evaluating soldiers' injuries and formulating treatment strategies.

3. Materials and Methods

In this section, we first explain the overall process of the research, then introduce the WTSS algorithm in detail. Next, we introduce the structure of our DNN classifier, and then determine the multiclassification metrics used in the algorithm to evaluate the performance of the classifier. Finally, the method of judging the plausibility of the generated synthetic samples is introduced.

3.1. Workflow of the Study

To solve the data augmentation problem and the supervised learning problem, an integrated modeling approach that incorporates the war trauma severity scoring algorithm (WTSS) and a DNN model was proposed. This approach's workflow is summarized as follows (Figure 1).

1. Based on the known probability distributions, the injured parts, injury types, and complications were randomly sampled and then combined to form a complete war trauma injury condition. Next, we used the WTSS algorithm to calculate the severity score and evaluate the consequences, after which the injury consequence label was marked.
2. After the data preprocessing, to test the accuracy of the injury consequence prediction, we trained a DNN classifier with the generated data and tested it with real data.
3. Through the Delphi method, the expert panel reached a consensus on unreasonable multiple injuries based on the domain knowledge [44] and then filtered out the unreasonable synthetic samples after the data generation.
4. After the predicted accuracy was evaluated and the unreasonable synthetic samples were filtered out, credible virtual war trauma data were finally output.

3.2. Random Injury Generation

In the injury generation process, we first randomly sampled the injured part according to the probability of occurrence; then, we randomly selected the possible injury types according to the injured part; finally, we randomly sampled whether it is accompanied by complications; if there were complications, we randomly selected the possible complications.

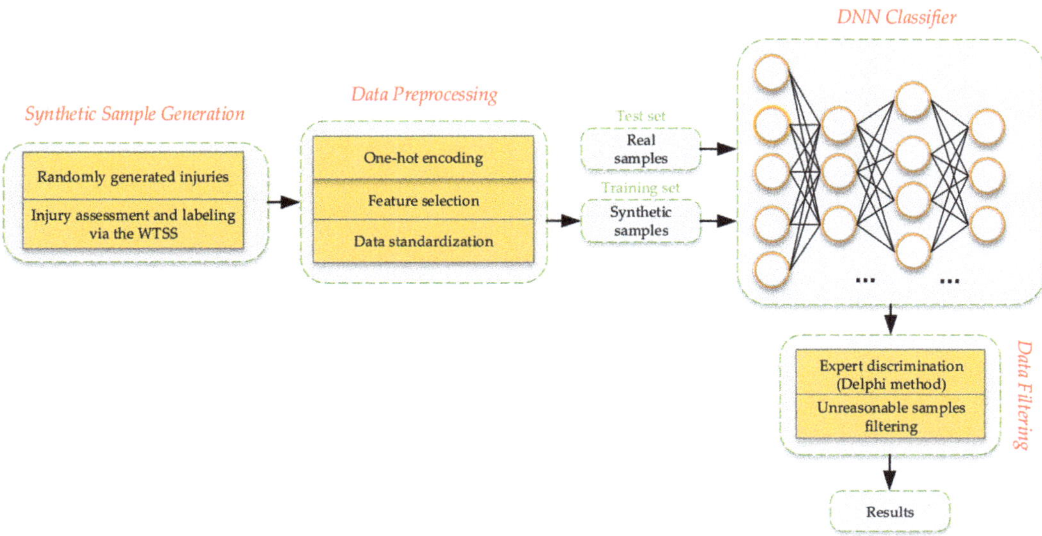

Figure 1. Workflow of the WTSS–DNN integrated approach.

3.3. WTSS Algorithm

After injuries were randomly generated, the focus of the research was on how to conduct standardized and accurate injury assessments. To solve this problem, we conducted multiple rounds of discussions and communication with the expert panel and finally decided to carry out a standardized quantitative assessment of various injuries by proposing a war trauma severity scoring algorithm.

Via in-depth summary of the various existing trauma scoring algorithms and based on the idea of multiple nonlinear regression and the key factors that affect severity of an injury (injured part, injury type, complications, and whether there are multiple injuries), after several rounds of testing and optimization, the equation for WTSS was finally determined as follows:

$$F(P, X, C) = a + \sum_{i=0}^{6} P_i X_i + C_i \tag{2}$$

where F represents the severity score; P_i represents the weight coefficient of injury severity for each of the seven body parts; X_i shows whether the corresponding body part was injured (if not injured, the corresponding X_i value equals 0; otherwise, it equals the injury severity standard score for the corresponding body part); C_i shows whether the injury was accompanied by complications (if there were no complications, C_i equals 0; otherwise, it equals the corresponding severity score); the bias a is the correction value for multiple injuries (if there were multiple injuries, a equals −20; otherwise, it equals 0).

Next, we calculated F according to the predictive factors P_i, X_i, C_i, and a, then selected the corresponding score interval according to the magnitude of F. Finally, we labeled the synthetic samples with the consequences of the injury. The pseudocode of WTSS is provided in Algorithm 1.

Algorithm 1. War trauma severity score (WTSS).

Input: Weight coefficient of injury parts: $P_i = \{P_0, P_1, \ldots, P_6\}$.
Injury type score: $X_i = \{X_0, X_1, \ldots, X_6\}$.
Complication score: $C_i = \{C_0, C_1, \ldots, C_6\}$.
Correction value for multiple injuries: $a = -20$.
Output: Severity score: $F(P, X, C)$.

```
n = 0
for i = 0 to 6 do
    if P_i ≠ 0 and X_i ≠ 0 then
        F(P, X, C) += P_i*X_i
        n += 1
    end if
    if C_i ≠ 0 then
        F(P, X, C) += C_i
    end if
end for
if n > 1 then
    F(P, X, C) += a
end if
return F(P, X, C)
```

The WTSS algorithm is a nonlinear model which ignores complicated details of the injury and uses a good correlation between the injuries' consequences and the severity of the injured parts and the injury types [45]. The weight coefficients of injuries in different body parts are shown in Table 1, and the example of the standard severity score for injury types and complications are shown in Figures 2 and 3. The score intervals for the injury consequences are listed in Table 2.

Table 1. Weight coefficients of each body part.

Body Part	Weight Coefficient
Head	8
Face	8
Neck	8
Chest / back	7
Abdomen	6.5
Pelvis / hip	6.5
Limbs	5

In a situation wherein different injury types or complications have the same standard injury severity score in a certain injured part, we coded them to distinguish. Taking the abdomen as an example, the coding method is shown in Figure 4.

As an independent scoring algorithm to determine severity of war trauma, WTSS does not perform an extremely accurate diagnosis of a specific injury. Instead, it performs standardized assessment and prediction of the most probable consequences of injuries from an objective perspective to ensure accuracy of the injury consequence assessment. Additionally, WTSS is not only the core of our WTSS–DNN integrated model that contributes to large-scale analysis and evaluation of war trauma data, but it also helps to quickly evaluate and diagnose soldiers' injuries on the battlefield and determine the treatment strategy. Furthermore, in complex battlefield environments, the soldier's age, physical constitution, and other factors may cause different consequences of the same trauma. Consequently, WTSS only objectively assesses the injury without considering the age and other physiological indicators to meet the requirements of the ideal scoring method that is "easy to implement, objective, and accurate" [38].

Injury part	Injury type (Score)			
Head	Perforating injury I (8) Perforating injury II (11)	Penetrating injury I (3.5) Penetrating injury II (10)	Fracture I (3) Fracture II (3)	
Face	Perforating injury I (8) Perforating injury II (7)	Penetrating injury I (6) Penetrating injury II (6)	Fracture I (3) Fracture II (3)	
Neck	Penetrating injury II (7)	Soft tissue injury I (5)	Soft tissue injury II (4)	
Chest / Back	Perforating injury I (7) Perforating injury II (8.5)	Penetrating injury I (3) Penetrating injury II (8)	Severe fracture I (7.5)	
Abdomen	Perforating injury I (7) Perforating injury II (7)	Penetrating injury I (8) Penetrating injury II (7.5)	Tangential injury I (3)	
Pelvis / Hip	Perforating injury I (9) Perforating injury II (8)	Penetrating injury I (8) Penetrating injury II (7.5)		
Limbs	Perforating injury I (7) Perforating injury II (5)	Penetrating injury I (3) Penetrating injury II (3)	Limb mutilation I (6.5) Limb mutilation II (6)	Tangential injury I (3) Tangential injury II (2)

Figure 2. Standard severity score for different injury types. In this Figure, I indicate that the injury is a blast injury, II indicates that the injury is a gunshot wound.

Injury part	Complication (Score)		
Head	Major bleeding caused by a penetrating injury (20) Major bleeding caused by a severe fracture (30)	Brain hernia (10) Increased intracranial pressure (7)	
Face	Shock (15)	Blindness (15)	
Neck	Dyspnea (20)	Hemorrhagic shock (25)	
Chest / Back	Major bleeding caused by a penetrating injury (10) Major bleeding caused by a perforating injury (15) Hemopneumothorax (30)	Hemorrhagic shock (25) Hemothorax shock (25) Cardiac rupture (40)	Pneumothorax (5) Hemothorax (20)
Abdomen	Major bleeding caused by a penetrating injury (15) Hemorrhagic shock (25)	Shock (10) Peritonitis (3)	
Pelvis / Hip	Shock (15)	Hemorrhagic shock (20)	
Limbs	Shock (15)	Hemorrhagic shock (25)	Major bleeding (15)

Figure 3. Standard severity score for different complications.

Table 2. Description of the score intervals.

Score Interval	Consequence	Label
0–25	Minor injury	1
26–50	Moderate injury	2
51–75	Serious injury	3
75+	Critical injury and death	4

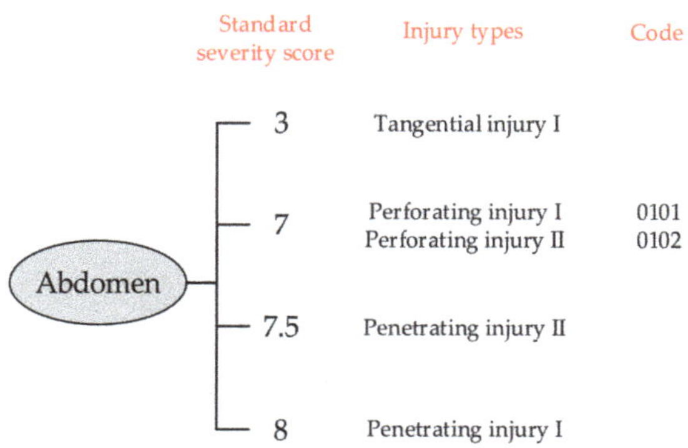

Figure 4. Example of injury coding in the abdominal area.

3.4. Deep Neural Network

Because the WTSS algorithm is a complicated nonlinear model, this article used a DNN as a classifier model to test the accuracy of injury consequences. The DNN classifier consists of an input layer, an output layer, and several hidden layers. It uses multilayer nonlinear information processing, which can be widely and flexibly used to solve problems such as classification, regression, dimensionality reduction, feature extraction, and clustering. First, we built a suitable DNN classifier network structure according to the actual needs, and the network structure was determined to be 22–16–16–16–4 after the experiment. Next, to test whether such a classifier has excellent generalization ability, we trained it with synthetic samples and tested it with real samples. To verify its performance, we used four multiclassification metrics based on a confusion matrix: accuracy, precision, recall, and the F_1 score [46]. Among these metrics, the F_1 score is the harmonic average of precision and recall. Finally, we adjusted and optimized the hyperparameters and then determined the best learning rate and the training sample size. The confusion matrix is shown in Figure 5.

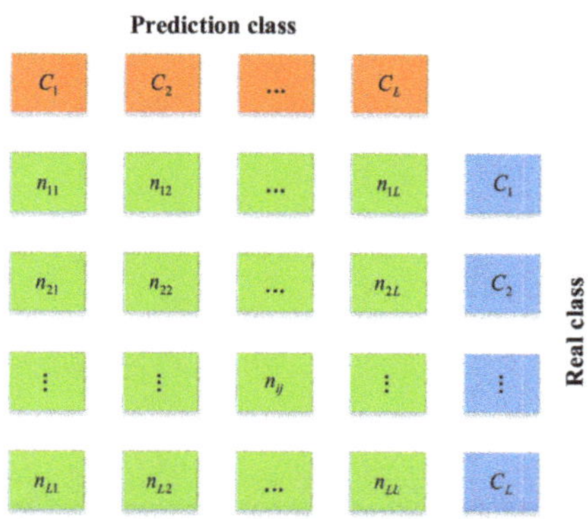

Figure 5. Graph of the multiclassification confusion matrix.

In Figure 5, L represents the class number, n_{ii} and n_{ij}—the number of class C_i samples correctly predicted as class C_i and incorrectly predicted as class C_i, respectively; R_i and P_i indicate the recall and the precision of class C_i, defined in Equations (3) and (4), and the accuracy and the F_1 score are defined in Equations (5) and (6).

$$P_i = \frac{n_{ii}}{\sum_{j=1}^{L} n_{ji}} \tag{3}$$

$$R_i = \frac{n_{ii}}{\sum\limits_{j=1}^{L} n_{ij}} \quad (4)$$

$$Accuracy = \frac{\sum\limits_{i=1}^{L} n_{ii}}{\sum\limits_{i=1,j=1}^{L} n_{ij}} \quad (5)$$

$$F_1\ score = 2\frac{\sum\limits_{i=1}^{L} R_i \sum\limits_{i=1}^{L} P_i}{\sum\limits_{i=1}^{L} R_i + \sum\limits_{i=1}^{L} P_i} \quad (6)$$

3.5. Discrimination of Unreasonable Injuries Based on the Delphi Method

After data generation, to improve the data plausibility of the synthetic samples, the expert panel reached a consensus on multiple unreasonable injuries based on the domain knowledge and provided feedback. Based on this feedback, we analyzed the law of unreasonable injury combinations and filtered out the unreasonable synthetic samples to improve data plausibility. Finally, we outputted the credible synthetic samples.

4. Empirical Analysis

Due to the high confidentiality and difficulty of access to war trauma data, it is gradually attracting greater attention from the army, military academies, and related hospitals. To eliminate obstacles to related research, an efficient and credible data augmentation approach is urgently needed in order to support large-scale war trauma data research and war game deduction. Our proposed integrated model provides a new and feasible way to meet the real need for large-scale and automated generation of credible war trauma data.

4.1. Data Collection

In this study, we collected and organized two types of real war trauma data at a certain scale: data on gunshot wounds and blast injuries. We selected 338 cases (minor injury, 114 cases; moderate injury, 82 cases; serious injury, 74 cases; and critical injury and death, 68 cases) complete with the available data to form the test set. After the preprocessing operations such as one-hot encoding, data standardization, and feature reduction, our war trauma data had a total of 22 features.

4.2. Results Analysis

We implemented our proposed WTSS–DNN integrated model in Python 3.7.7 and conducted experiments on a personal computer with a Windows 64-bit operating system. After a series of tests on the DNN, the optimal values of all the hyperparameters were determined. The classifier's input dimension was 22, equal to the feature dimension of the war trauma samples. The number of hidden layers of the classifier was set at 4, with each using ReLUs as the activation function. The softmax function was used as the output layer, and categorical cross-entropy was used as the loss function. We used TensorFlow 2.0.0 and GPU to train our DNN classifier; the epoch was set at 1000 and the batch size was set at 256. We chose Adam as our optimization algorithm as it performed best compared to SGD and RMSProp3 [47].

After determining the best network structure of the DNN classifier (22–16–16–16–16–4), we conducted contrast experiments at different learning rates [48]. Specifically, we kept the network structure and other hyperparameters unchanged, then set the values of the learning rate to be 0.05, 0.02, 0.01, 0.005, 0.002, 0.001, 0.0005, and 0.0001, respectively. Table 3 shows accuracy, precision, recall, and the F_1 score at different learning rates on the same training set with a sample size of 10,000. The results show that the 0.001 learning rate led to the best overall model performance and thus was selected and used.

Table 3. Comparison of the multiclassification metrics at different learning rates.

LR	Accuracy	Precision	Recall	F_1 Score
0.05	48.82	48.89	51.11	49.98
0.02	49.41	49.47	51.85	50.63
0.01	59.99	67.35	66.59	66.97
0.005	65.43	72.51	71.15	71.82
0.002	73.19	78.46	80.37	79.40
0.001	81.57	88.08	87.70	87.89
0.0005	78.33	86.33	85.74	86.03
0.0001	73.37	79.57	80.85	80.20

Next, we explored the best training sample size (n). On the one hand, low numbers of training samples cannot fully teach sample features and meet the requirements of model accuracy; and on the other hand, too high numbers of training samples can increase the calculation costs and time costs and are not conducive to optimizing the hyperparameters. Therefore, we sought to determine the best training sample size in the range of 1000–20,000 through the trial and error method [49]. In the search process, to avoid the impact of class imbalance on the experimental results, synthetic samples of the four classes were extracted at the same proportion to form a training set for the experiment and test. The overall performance results of the multiclassification metrics at different training sample sizes are shown in Table 4.

Table 4. Comparison of the multiclassification metrics at different sample sizes.

n	Accuracy	Precision	Recall	F_1 Score
1000	73.17	78.51	79.00	78.75
2000	75.15	81.67	80.96	81.31
4000	78.99	85.14	84.70	84.91
8000	80.88	87.97	86.81	87.39
12,000	84.33	90.07	88.44	89.25
16,000	82.36	89.07	89.10	89.08
20,000	80.18	87.21	88.33	87.77

The experimental results showed that the small-scale training set did not meet the requirements for model accuracy. As the training sample size continued to increase, the predicted accuracy gradually increased. When the training sample size was 8000, the accuracy reached 80.88%; and when the training scale increased to 12,000, the accuracy increased to 84.33%. However, model performance became deteriorated when the training scale was greater than 12,000, which indicates that blindly increasing the training scale could not guarantee a consistently higher classification accuracy. Besides, when the training scale was increased, as the harmonic average of precision and recall, the trend of the F_1 score was basically consistent with that of accuracy. Therefore, we supposed that selecting a training sample size of 12,000 can achieve the best compromise between the training cost and the classification performance.

Finally, our DNN classifier achieved the best overall performance with 84.33% accuracy, 90.07% precision, 88.44% recall, and an 89.25% F_1 score.

4.3. Evaluation of WTSS Combined with a DNN

In this section, we first explored the accuracy of injury assessment of different classifier models. Subsequently, to evaluate the respective contributions of the WTSS algorithm and the DNN classifier in the WTSS–DNN data augmentation method, we set up an ablation experiment. Finally, we provided the prediction results of the DNN for real data through the confusion matrix.

First, we compared our DNN model with three classic machine-learning classifiers: random forest (RF) [50], XGBoost [51], and naïve Bayes (NB) [52].

The RF, XGBoost, and NB models and our DNN model were trained with the same training set and then tested with the same real samples. As shown in Figure 6, our DNN classifier performed better than the three classic machine-learning models. The NB model showed the weakest performance in comparison with the other classifier models because when the number of features is large or when the correlation between the features is high, the NB classification effect is poor. These results indicate that classic machine-learning models cannot be effectively trained when there are few samples and verified that a DNN classifier trained with a large amount of data has better classification performance.

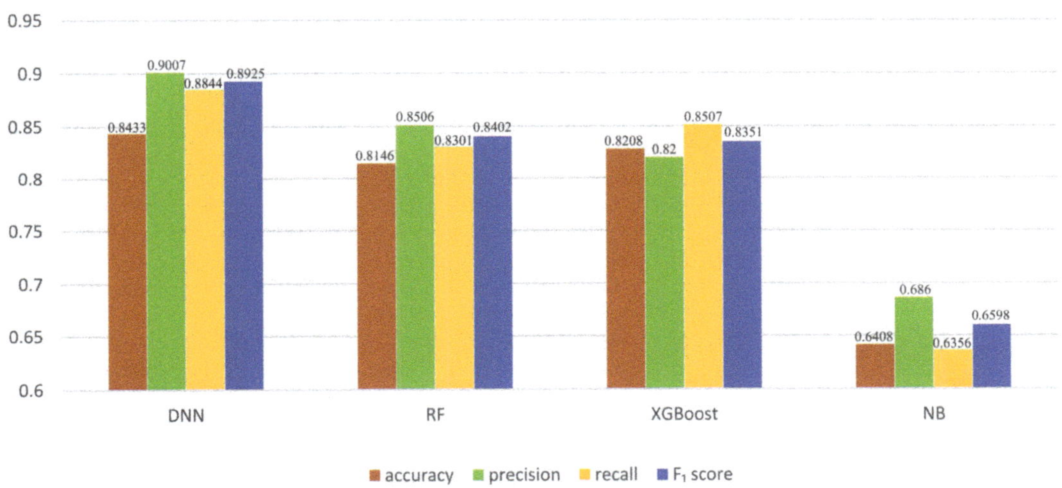

Figure 6. Performance of the different classification strategies.

Next, to evaluate the respective contributions of the WTSS algorithm and the DNN classifier in the WTSS–DNN integrated model, we set up an ablation experiment. Specifically, we combined different injury assessment methods with different classifier models to observe performance of various combinations. Injury assessment methods include the WTSS algorithm and the manual assessment method (MA); classification models include DNN, RF, XGBoost, and NB. The results of the ablation experiment are shown in Table 5.

Table 5. Ablation experiment of different injury assessment methods and classifier models.

Assessment Method	Classifier Model	Accuracy	Precision	Recall	F_1 Score
MA	RF	69.39	70.21	70.47	70.34
	XGBoost	68.86	69.12	70.07	69.59
	NB	53.86	55.20	53.41	54.29
	DNN	71.24	72.66	70.81	71.72
WTSS	RF	81.46	85.06	83.01	84.02
	XGBoost	82.08	82.00	85.07	83.51
	NB	64.08	68.60	63.56	65.98
	DNN	84.33	90.07	88.44	89.25

From the results of the ablation experiment, we can see that the WTSS algorithm is better than the traditional manual evaluation method, the prediction performance of the DNN classifier is better than that of the machine-learning model, and the combination of WTSS and the DNN performs best. Therefore, the combination of WTSS and the DNN can effectively solve the data augmentation problem of war trauma data and shows superiority compared with artificial generation methods.

Finally, we provided the prediction results of the DNN for real data through the confusion matrix.

From Table 6, we can see that the prediction accuracy for minor injuries and moderate injuries is very high, but the prediction accuracy for critical injuries is only about 60%, which is caused by the complexity of critical injuries.

Table 6. Confusion matrix of injury consequence identification.

Real \ Predicted	Minor Injury	Moderate Injury	Serious Injury	Critical Injury and Death	Total
Minor injury	114	0	0	0	114
Moderate injury	3	73	6	0	82
Serious injury	0	7	60	7	74
Critical injury and death	0	10	20	38	68
Total	117	90	86	45	338

4.4. Data Filtering

The Delphi method, also known as the "expert investigation method", was invented in 1946 by RAND Corporation in the United States. The Delphi method is based on the key assumption that predictions from groups are usually more accurate than predictions from individuals. The goal of this method is to use a structured iterative approach to obtain consensual opinions from an expert panel [44].

For the multiple injuries data generated, some injury combinations are unreasonable—they are almost impossible to appear in a real war. To improve plausibility and usability of the synthetic samples in our experiment, we decided to use the Delphi method to evaluate unreasonable multiple injuries and filter them out. After several rounds of identification and discussions, the expert panel reached a consensus on the unreasonable multiple injuries based on the domain knowledge. We analyzed the experts' feedback and then filtered out the unreasonable synthetic samples to improve data plausibility to output credible samples. Next, to verify whether the data plausibility improved or not, we randomly selected 300 original multiple-injury synthetic samples and 300 filtered ones, put them into three groups, and conducted contrast experiments. Then, we counted the number of reasonable samples before and after filtering. The experimental results are shown in Figure 7.

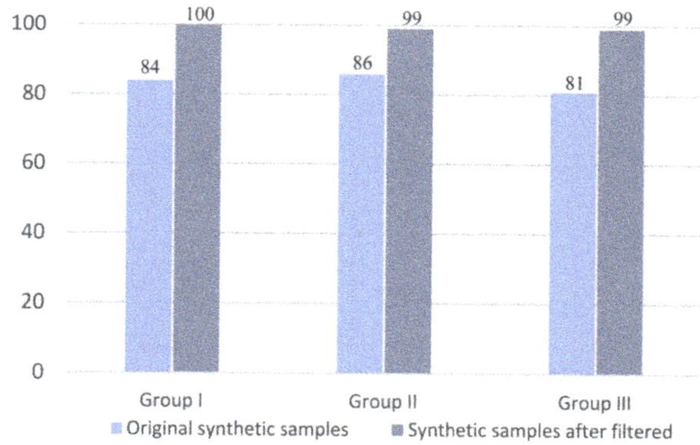

Figure 7. The numbers of reasonable samples in the synthetic samples.

The experimental results showed that data plausibility of the synthetic samples filtered out was significantly improved in comparison with that of the original ones and came close to 100%.

5. Discussion

For the WTSS–DNN integrated model, plausibility and effectiveness of the WTSS algorithm play a crucial role in the performance of WTSS–DNN. Therefore, we evaluated plausibility and effectiveness of the WTSS algorithm through the two methods described below. First, the expert panel intervention and assistance. The parameter setting and the scoring standard of the algorithm were determined after multiple rounds of discussions and evaluations with the expert panel, which is highly reasonable and professional. Second, we tested plausibility and effectiveness of the algorithm through ablation experiments. In the ablation experiments, on the one hand, we used the DNN classifier to verify accuracy and plausibility of the algorithm in injury assessment. The experimental results show that the prediction accuracy rate reached 84.43%, which is a satisfactory result. On the other hand, we compared the WTSS algorithm with the traditional manual assessment method, further verified plausibility and superiority of the WTSS algorithm in injury assessment. Therefore, compared with the artificially generated methods, the performance of the proposed WTSS algorithm combined with a DNN in war trauma data augmentation is superior, can ensure high data quality, and automatically generates large-scale war trauma data on demand.

However, the experiment also showed that the prediction accuracy of the severity of multiple injuries was lower than that for a single injury due to the complexity of multiple injuries. Furthermore, after determining the WTSS standards, the proposed approach no longer relies on additional professional knowledge due to the characteristics of DL. Thus, for nonprofessionals, the proposed approach has a low barrier to successful application. Although we were able to generate credible virtual trauma data only for blast injuries and gunshot wounds in this study, with the continuous real data collection, the types of war trauma we can generate will become more abundant. Finally, the combination of DL with medical scoring algorithms can be used for other types of injury data augmentation, such as for surgical injuries and emergency injuries.

6. Conclusions

In this article, the WTSS algorithm combined with a DNN was presented for the augmentation of war trauma data. Compared with the traditional artificial data augmentation method, our integrated modeling approach not only improves the quality of injury consequence assessment, but can also automatically generate large-scale and credible virtual war trauma data. The generated data make it possible to carry out related data-based military research, which has great practical significance and value. In addition, it also provides a practical auxiliary tool for quickly evaluating soldiers' injuries and formulating treatment strategies, which are of crucial significance to the analysis and evaluation of war trauma data. Finally, because this study was the first attempt to combine DL and the trauma scoring algorithm for the augmentation of war trauma data, it still had some shortcomings, but with the continuous improvement of the WTSS algorithm, the performance of our WTSS–DNN integrated model will become more superior. That is also the focus and direction of our future research, to continuously improve the comprehensiveness and applicability of our integrated modeling approach.

Author Contributions: P.Z. conceived the presented idea and verified the analytical methods. J.Y. provided the experimental environment, supervised and validated the findings of this work. Y.Z. provided the research topic, medical theoretical and technical support. Writing, editing, and formatting the manuscript was carried out by P.Z. with support from Y.Z. and J.Y. Funding acquisition was carried out by S.W. and Y.H. All authors have read and agreed to the published version of the manuscript.

Funding: This work was supported by the National Natural Science Fund, sponsor: Jibin Yin, funding number: 61741206.

Acknowledgments: We would like to thank Shuoyu Wang and Yi Han for their assistance with this study.

Conflicts of Interest: The authors declare no conflict of interest.

References

1. Cardi, M.; Ibrahim, K.; Alizai, S. Injury patterns and causes of death in 953 patients with penetrating abdominal war wounds in a civilian independent non-governmental organization hospital in Lashkargah, Afghanistan. *World J. Emerg. Surg.* **2019**, *14*, 1–8. [CrossRef] [PubMed]
2. Crevier, D. *AI: The Tumultuous History of the Search for Artificial Intelligence*; Basic Books, Inc.: New York, NY, USA, 1993.
3. Pan, Y. Heading toward Artificial Intelligence 2.0. *Engineering* **2016**, *2*, 409–413. [CrossRef]
4. Zhuang, Y.; Chen, C.; Wu, F.; Pan, Y. Challenges and opportunities: From big data to knowledge in AI 2.0. *Front. Inf. Technol. Electron. Eng.* **2017**, *18*, 3–14. [CrossRef]
5. Salem, A.; Moselhi, O. AI-based cloud computing application for smart earthmoving operations. *Can. J. Civ. Eng.* **2021**, *48*, 312–327. [CrossRef]
6. Zheng, A.; Chen, Z.; Li, C. Learning Deep RGBT Representations for Robust Person Re-identification. *Int. J. Autom. Comput.* **2021**, *18*, 443–456. [CrossRef]
7. Liu, J.; Wang, L.; Zhou, H. The Application of Human–Computer Interaction Technology Fused with Artificial Intelligence in Sports Moving Target Detection Education for College Athlete. *Front. Psychol.* **2021**, 2848. [CrossRef]
8. Tang, Z.; Shao, K.; Zhao, D.; Zhu, Y. Recent progress of deep reinforcement learning: From AlphaGo to AlphaGo Zero. *Control Theory Appl.* **2017**, *34*, 1529–1546. [CrossRef]
9. He, T.; Mamta, P.; Richard, O.; James, M.; Yu, X.; Chen, S. Deep learning analytics for diagnostic support of breast cancer disease management. In Proceedings of the IEEE EMBS International Conference on Biomedical & Health Informatics (BHI), Orlando, FL, USA, 16–19 February 2017.
10. Duan, X.; Yang, Y.; Tan, S.; Wang, S.; Feng, X.; Cui, L.; Feng, F.; Yu, S.; Wang, W.; Wu, Y. Application of artificial neural network model combined with four biomarkers in auxiliary diagnosis of lung cancer. *Med Biol. Eng. Comput.* **2017**, *55*, 1239–1248. [CrossRef]
11. King, P. Signal Processing and Machine Learning for Biomedical Big Data. *IEEE Pulse* **2019**, *10*, 34–35. [CrossRef]
12. Lee, J.; Jun, S.; Cho, Y.; Lee, H.; Kim, G.; Seo, J.; Kim, N. Deep learning in medical imaging: General overview. *Korean J. Radiol.* **2017**, *18*, 570–584. [CrossRef]
13. Fan, R.; Zhang, N.; Yang, L.; Ke, J.; Zhao, D.; Cui, Q. AI-based prediction for the risk of coronary heart disease among patients with type 2 diabetes mellitus. *Sci. Rep.* **2020**, *10*, 14457. [CrossRef]
14. Rong, G.; Mendez, A.; Assi, E.; Zhao, B.; Sawan, M. Artificial Intelligence in Healthcare: Review and Prediction Case Studies. *Engineering* **2020**, *6*, 91–301. [CrossRef]
15. Menchón-Lara, R.; Sancho-Gómez, J.; Bueno-Crespo, A. Early-stage atherosclerosis detection using deep learning over carotid ultrasound images. *Appl. Soft Comput.* **2016**, *49*, 616–628. [CrossRef]
16. Liu, Y.; Zhou, Y.; Liu, X.; Wang, C.; Wang, Z. Wasserstein GAN-Based Small-Sample Augmentation for New-Generation Artificial Intelligence: A Case Study of Cancer-Staging Data in Biology. *Engineering* **2019**, *5*, 156–163. [CrossRef]
17. Ellery, W.; David, F. Deep Learning-Based Survival Prediction for Multiple Cancer Types Using Histopathology Images. *PLoS ONE.* **2020**, *15*, e0233678. [CrossRef]
18. Joohi, C.; Puneet, G. BPBSAM: Body part-specific burn severity assessment model. *Burns* **2020**, *46*, 1407–1423. [CrossRef]
19. Goodfellow, I.; Bengio, Y.; Courville, A. *Deep Learning*; MIT Press: Cambridge, MA, USA, 2016; p. 226.
20. Mamoshina, P.; Vieira, A.; Putin, E.; Zhavoronkov, A. Applications of deep learning in biomedicine. *Mol. Pharm.* **2016**, *13*, 1445–1454. [CrossRef]
21. Miotto, R.; Wang, F.; Wang, S.; Jiang, X.; Dudley, J. Deep learning for healthcare: Review, opportunities and challenges. *Brief. Bioinform.* **2018**, *19*, 1236–1246. [CrossRef] [PubMed]
22. Bozkurt, S.; Gimenez, F.; Burnside, E. Using automatically extracted information from mammography reports for decision-support. *J. Biomed. Inform.* **2016**, *62*, 224–231. [CrossRef]
23. DeRouin, E.; Brown, J.; Beck, H.; Fausett, L.; Schneider, M. Neural network training on unequally represented classes. *Intell. Eng. Syst. Artif. Neural Netw.* **1991**, *1*, 135–145.
24. Chawla, V.; Bowyer, W.; Hall, L. SMOTE: Synthetic minority over-sampling technique. *J. Artif. Intell. Res.* **2002**, *16*, 21–357. [CrossRef]
25. Han, H.; Wang, W.; Mao, B. Borderline-SMOTE: A new over-sampling method in imbalanced data sets learning. In Proceedings of the International Conference on Intelligent Computing, Hefei, China, 23–26 August 2005; Springer: Berlin/Heidelberg, Germany, 2005; pp. 878–887.
26. Cieslak, A.; Chawla, V.; Striegel, A. Combating imbalance in network intrusion datasets. In Proceedings of the 2006 IEEE International Conference on Granular Computing, Atlanta, GA, USA, 10–12 May 2006; pp. 732–737.

27. He, H.; Bai, Y.; Garcia, A.; Li, S. ADASYN: Adaptive synthetic sampling approach for imbalanced learning. In Proceedings of the 2008 IEEE International Joint Conference on Neural Networks (IEEE World Congress on Computational Intelligence), Hong Kong, China, 1–8 June 2008; pp. 1322–1328. [CrossRef]
28. Barua, S.; Islam, M.; Yao, X.; Murase, K. MWMOTE–Majority weighted minority oversampling technique for imbalanced data set learning. *IEEE Trans. Knowl. Data Eng.* **2012**, *26*, 405–425. [CrossRef]
29. Douzas, G.; Bacao, F. Self-Organizing Map Oversampling (SOMO) for imbalanced data set learning. *Expert Syst. Appl.* **2017**, *82*, 40–52. [CrossRef]
30. Goodfellow, I.; Pouget-Abadie, J.; Mirza, M.; Xu, B.; Warde-Farley, D.; Ozair, S. Generative adversarial nets. *Adv. Neural Inf. Process. Syst.* **2014**, *2*, 27.
31. Wang, K.; Gou, C.; Duan, Y.; Lin, Y.; Zheng, X.; Wang, F. Generative adversarial networks: Introduction and outlook. *IEEE/CAA J. Autom. Sin.* **2017**, *4*, 588–598. [CrossRef]
32. Creswell, A.; White, T.; Dumoulin, V.; Arulkumaran, K.; Sengupta, B.; Bharath, A. Generative adversarial networks: An overview. *IEEE Signal Process. Mag.* **2018**, *35*, 53–65. [CrossRef]
33. Canzi, G.; De Ponti, E.; Novelli, G.; Mazzoleni, F.; Chiara, O.; Bozzetti, A.; Sozzi, D. The CFI score: Validation of a new comprehensive severity scoring system for facial injuries. *J. Cranio-Maxillofac. Surg.* **2019**, *47*, 377–382. [CrossRef]
34. Ratnayake, H.; Johnson, D.; Martensson, J.; Lam, Q.; Bellomo, R. A laboratory-derived early warning score for the prediction of in-hospital mortality, ICU admission, Medical Emergency Team activation and Cardiac Arrest in general medical wards. *Intern. Med. J.* **2019**, 37. [CrossRef]
35. Trongtrakul, K.; Patumanond, J.; Kongsayreepong, S.; Morakul, S.; Pipanmekaporn, T.; Akaraborworn, O.; Poopipatpab, S. Acute kidney injury risk prediction score for critically-ill surgical patients. *BMC Anesthesiol.* **2020**, *20*, 1–10. [CrossRef]
36. Zhou, J. Introduction of Trauma Scoreology. *Inj. Med.* **2018**, *7*, 4–9.
37. de Alencar Domingues, C.; Coimbra, R.; Sérgio, R.; Poggetti, R.; de Souza Nogueira, L.; de Sousa, R. New Trauma and Injury Severity Score (TRISS) adjustments for survival prediction. *World J. Emerg. Surg.* **2018**, *13*, 12. [CrossRef]
38. Yang, J.; Liu, Z.; Zhang, Y. Comparison of injury degree score method and modified trauma severity score method for inpatients with accidental injury. *Med. J. Natl. Defending Forces Northwest China* **2017**, *38*, 364–368.
39. Shi, J.; Shen, J.; Zhu, M.; Wheeler, K.; Lu, B.; Kenney, B.; Xiang, H. A new weighted injury severity scoring system: Better predictive power for adult trauma mortality. *Inj. Epidemiol.* **2019**, *6*, 40. [CrossRef] [PubMed]
40. Wang, H.; Liu, X.; Wang, G. Application and analysis of revised injury severity score on emergencies. *Orthop. Biomech. Mater. Clin. Study* **2014**, *11*, 20–22.
41. Yang, T.; Guo, P.; Liu, W.; Liu, X.; Hao, T. Enhancing PIR-based Multi-person Localization through Combining Deep Learning with Domain Knowledge. *IEEE Sens. J.* **2020**, 1. [CrossRef]
42. Ding, L.; He, H. Research on domain adaptation of machine translation based on domain knowledge and deep learning. *Inf. Sci.* **2017**, *35*, 125–132.
43. Mcdaniel, P.; Papernot, N.; Celik, B. Machine learning in adversarial settings. *IEEE Secur. Priv.* **2016**, *14*, 68–72. [CrossRef]
44. Gordon, J. The delphi method. *Futures Res. Methodol.* **1994**, *2*, 1–30.
45. Penn-Barwell, J.G.; Bishop, R.; Midwinter, J. Refining the Trauma and Injury Severity Score (TRISS) to Measure the Performance of the UK Combat Casualty Care System. *Mil. Med.* **2018**, *183*, e442–e447. [CrossRef]
46. Liu, Y.; Xu, Z.; He, J.; Wang, Q. Data Augmentation Method for Power Transformer Fault Diagnosis Based on Conditional Wasserstein Generative Adversarial Network. *Power Syst. Technol.* **2020**, *44*, 1505–1513.
47. Diederik, K. Adam: A Method for Stochastic Optimization. In Proceedings of the 3rd International Conference on Learning Representations, Banff, AB, Canada, 14–16 April 2014.
48. Zhang, H.; Huang, L.; Wu, C.; Li, Z. An effective convolutional neural network based on SMOTE and Gaussian mixture model for intrusion detection in imbalanced dataset. *Comput. Netw.* **2020**, 177. [CrossRef]
49. Xu, R.; Cao, J.; Wu, Y.; Wang, S.; Luo, J.; Chen, X.; Fang, F. An integrated approach based on virtual data augmentation and deep neural networks modeling for VFA production prediction in anaerobic fermentation process. *Water Res.* **2020**, *184*, 116103. [CrossRef] [PubMed]
50. Breiman, L. Random Forests. *Mach. Learn.* **2001**, *45*, 5–32. [CrossRef]
51. Chen, T.; Guestrin, C. XGBoost: A Scalable Tree Boosting System. In Proceedings of the 22nd ACM SIGKDD International Conference on Knowledge Discovery and Data Mining, San Francisco, CA, USA, 13–17 August 2016; pp. 785–794.
52. Webb, G. Naïve Bayes. *Encycl. Mach. Learn.* **2010**, *15*, 713–714.

electronics

Article

Research on fNIRS Recognition Method of Upper Limb Movement Intention

Chunguang Li [1,2], Yongliang Xu [1,2,*], Liujin He [1,2], Yue Zhu [1,2], Shaolong Kuang [1,2] and Lining Sun [1,2,*]

1 Key Laboratory of Robotics and System of Jiangsu Province, School of Mechanical and Electric Engineering, Soochow University, Suzhou 215000, China; lichunguang@suda.edu.cn (C.L.); 20175229083@stu.suda.edu.cn (L.H.); xcyd_zy@163.com (Y.Z.); kuangshaolong@gmail.com (S.K.)
2 Collaborative Innovation Center of Suzhou Nano Science and Technology, Suzhou 215000, China
* Correspondence: 20205229077@stu.suda.edu.cn (Y.X.); Lnsun@hit.edu.cn (L.S.)

Abstract: This paper aims at realizing upper limb rehabilitation training by using an fNIRS-BCI system. This article mainly focuses on the analysis and research of the cerebral blood oxygen signal in the system, and gradually extends the analysis and recognition method of the movement intention in the cerebral blood oxygen signal to the actual brain-computer interface system. Fifty subjects completed four upper limb movement paradigms: Lifting-up, putting down, pulling back, and pushing forward. Then, their near-infrared data and movement trigger signals were collected. In terms of the recognition algorithm for detecting the initial intention of upper limb movements, gradient boosting tree (GBDT) and random forest (RF) were selected for classification experiments. Finally, RF classifier with better comprehensive indicators was selected as the final classification algorithm. The best offline recognition rate was 94.4% (151/160). The ReliefF algorithm based on distance measurement and the genetic algorithm proposed in the genetic theory were used to select features. In terms of upper limb motion state recognition algorithms, logistic regression (LR), support vector machine (SVM), naive Bayes (NB), and linear discriminant analysis (LDA) were selected for experiments. Kappa coefficient was used as the classification index to evaluate the performance of the classifier. Finally, SVM classification got the best performance, and the four-class recognition accuracy rate was 84.4%. The results show that RF and SVM can achieve high recognition accuracy in motion intentions and the upper limb rehabilitation system designed in this paper has great application significance.

Keywords: brain–computer interface; intent recognition; SVM; ensemble learning

Citation: Li, C.; Xu, Y.; He, L.; Zhu, Y.; Kuang, S.; Sun, L. Research on fNIRS Recognition Method of Upper Limb Movement Intention. *Electronics* **2021**, *10*, 1239. https://doi.org/10.3390/electronics10111239

Academic Editors: João Paulo Morais Ferreira and Tao Liu

Received: 3 April 2021
Accepted: 15 May 2021
Published: 24 May 2021

1. Introduction

The upper extremity is an important part of the human body. Research has found that 80% of severe stroke patients have upper extremity motor dysfunction. It is a relatively feasible and efficient treatment method to perform rehabilitation training by using rehabilitation robot equipment to drive patients. However, in the traditional rehabilitation robot training scheme, the robot usually assists the patient to complete the training action after the specific training process is set [1]. The form of this program is very simple, and patients may feel negative and slack during the training process due to boredom.

Many previous studies have shown that the process of autonomous training by patients is very important. Compared with passive exercise training, the active willingness of patients to participate in training can better promote neurocortical reconstruction and motor function recovery [2]. As a new human-computer interaction method, the brain-computer interface (BCI) can bypass the function of nerve transmission channels and muscle parts, and directly establish information communication channels between the brain and the external environment, and control external devices. The application of BCI in the field of rehabilitation has helped a lot of patients with limb dysfunction to carry out rehabilitation training and accelerate their rehabilitation process. Therefore, in the field of rehabilitation medicine, the study of feasible BCI technology has very important social significance [3–5].

At present, many BCI researchers at home and abroad have focused on applying BCI technology to the field of upper limb motor function rehabilitation, and have obtained excellent research results. Anirban of the University of Essex and his research partners successfully developed a hybrid BCI device to control the exoskeleton of the hand in order to overcome the problem of low recognition accuracy in BCI system. The system combines EEG and EMG signals. After the grasping intention of the subject is successfully detected, the exoskeleton will perform finger flexion and extension. Finally, the recognition accuracy of the system reached (90.00 $\pm$ 4.86)%, significantly improving the performance of BCI system [6]. Zhai Wenwen hoped to improve the life independence of patients with severe motor dysfunction through BCI technology. The upper-limb movement-related instructions can control the robotic arm to complete the rehabilitation training of the shoulder, wrist and elbow. The recognition accuracy of the system is as high as 93% [7]. Yoshiyuki Suzuki studied the effects of human corticospinal excitability on motor tasks in the process of imagining or observing the upper limbs. The experiments have shown that kinesthetic MI, including visualizing and observing the virtual hand, can cause phase-dependent muscle-specific corticospinal stimulation of wrist muscles that match those in the actual hand [8]. Although they have achieved remarkable research results in the field of sports rehabilitation technology, there are still many key technologies that need to be improved. For example, the recognition accuracy of multi-classification tasks is low, real-time performance needs to be improved, and it is difficult for users to autonomously control the pace of rehabilitation training.

This paper proposes a set of upper limb rehabilitation training robot system based on user spontaneous movement fNIRS-BCI. Four upper limb movement paradigms are designed: Lifting up, putting down, pulling back, and pushing forward. The start of each task and the rest time were all controlled by the subjects autonomously without any prompts from the outside world [9]. A variety of classifiers such as RF and SVM were selected for evaluation, and a high accuracy rate was achieved. Furthermore, the most suitable multiclass recognition algorithm was selected.

2. Experimental Design

2.1. Participants

A total of 50 volunteers were recruited for this experiment. Among them, 29 were male volunteers and 21 were female volunteers. The ratio of male to female was approximately 3:2. All volunteers were right-handed, in good health, and had no history of mental illness or cardiovascular and cerebrovascular diseases. In addition, all volunteers participated in the experiment for the first time and only once. Before the experiment, all volunteers were informed of the experiment details and signed the informed consent form of the experiment.

2.2. Experiment Paradigm

In order to increase the controllability and practicability of upper-limb auxiliary equipment, four common upper-limb movement paradigms in daily life designed in daily life: Lifting up, putting down, pulling back and pushing forward. During the experiment, the near-infrared data of the subjects in different motion states were collected to provide a supervised learning data set for subsequent research on motion intention recognition. The experimental process is shown in Figure 1. Tasks 1, 2, 3, and 4 in the figure represent the four action paradigms of lifting up, putting down, pulling back, and pushing forward, respectively.

During the experiment, the subjects should keep their scalp clean and keep their hair dry before the experiment. It should be noted that the subjects need to rest for about 60 s before the start of the experiment. The rest time between lifting and lowering actions was 10 s, and the rest time between other actions was about 50 s. The start of each task and the rest time were all controlled by the subjects autonomously without any prompts from the outside world. The specific paradigm of the experiment is shown in Figure 2. Figure 2a–d represents the four action paradigms of lifting up, putting down, pulling back, and pushing forward, respectively.

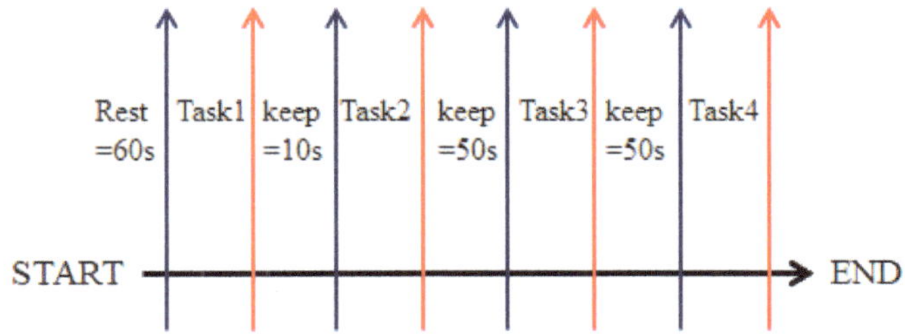

Figure 1. The diagram of the experiment.

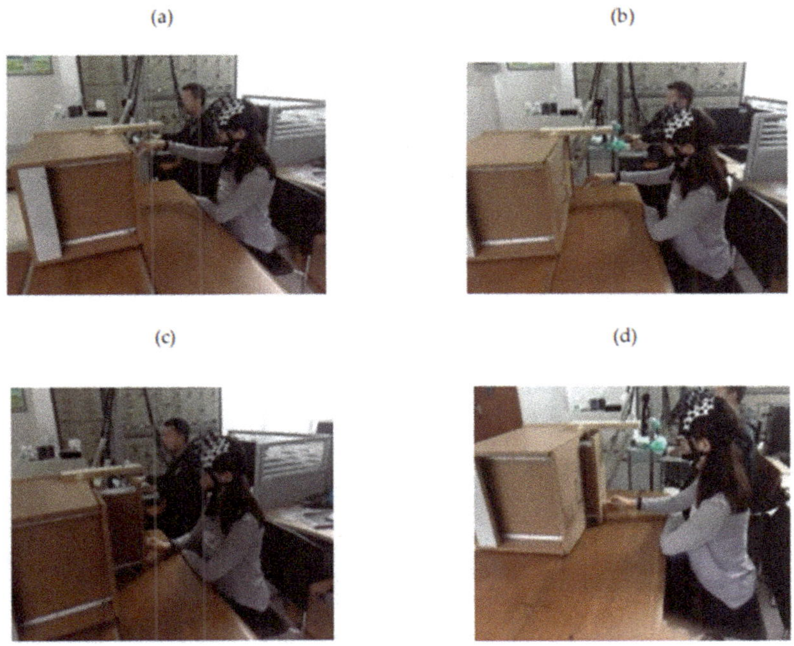

Figure 2. The experimental diagram of upper limb movement. (**a**) Lifting up, (**b**) putting down, (**c**) pulling back, and (**d**) pushing forward.

2.3. Cortical Regions

The FOIRE-3000 near-infrared acquisition system of Shimadzu Corporation was used as a brain signal acquisition device to record the changes in the concentration of oxygen, deoxygenation, and total oxygen hemoglobin in the experiment [10]. Because this experiment emphasized the spontaneous movement intention of the subjects, task execution and rest time were controlled by the subjects spontaneously, we chose to use continuous mode to measure the cerebral hemoglobin information during the experiment. The equipment sampling time was 130 ms.

This study used the internationally recognized 10–20 system as the positioning standard to locate brain functional areas. The experiment designed a 4 × 4 headgear layout to detect the above key functional areas. There were a total of 24 effective test channels. The overall headgear layout is shown in Figure 3. In the layout, the Cz point was used as a reference point for the relative position, and the distance between Cz and the center of the 23rd channel was 1.5 cm. The entire layout can cover four brain functional areas: PFC, SMA, PMC (PMC = PMCL + PMCR), PMCR, and M1 brain area function [11].

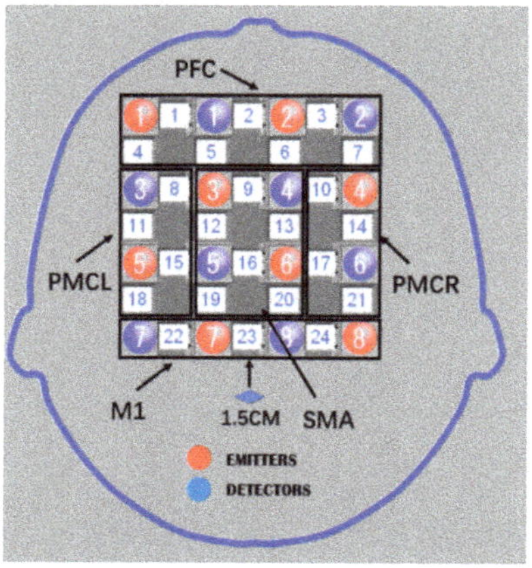

Figure 3. The schematic diagram of 4 × 4 headgear layout.

3. Data Analysis

3.1. Data Preprocessing

The interference components in the near-infrared signal mainly include baseline drift, physiological interference, and high-frequency noise. Mathematical morphology has a very strict data theoretical basis, and research shows that this method achieved good results in nonlinear signal processing. Corrosion calculation and expansion calculation were the basic calculation methods of this method. Based on the corrosion expansion calculation, two different morphological calculations were used for the combination of opening and closing. Among them, the open operation used the first expansion and then the corrosion operation method to eliminate the peak of the signal to filter the peak noise above the signal, and the closed operation used the first erosion and then the expansion operation to fill the signal trough to smooth or suppress the signal valley noise. We combined the on-off filter to eliminate both positive and negative impulse noises in the signal to avoid unidirectional deviation of the filtered signal [12,13].

3.2. Data Preprocessing

3.2.1. Action Initiation Intent Feature Extraction

In order to accurately and quickly detect the time point of the subject's transition from resting to exercise, the selected feature has the ability to detect sudden changes and singular signals. Near-infrared signals are non-stationary and nonlinear. Among a variety of feature categories based on biological information to distinguish resting/exercise status, Teager–Kaiser energy operator and slope value were used in many initial detection studies of biological signals [14,15]. By calculating the Teager–Kaiser energy operator at each time point and the blood oxygen slope characteristics at each time point, the instantaneous changes attributed to the signal at each time point were obtained, and the waveform changes of the measured signal were tracked in real time.

3.2.2. Motion State Feature Extraction

In the exercise state classification task, the goal was to recognize the exercise intention before the real action. Therefore, it was necessary to extract the data of the real pre-exercise time period for analysis. The training time period selection method is shown in Figure 4. Took the starting point of the real action as the reference point (0), took the signal from −3 s to 0 s as the training time period, and applied the sliding time window method. The length of the window was set to 20. The overlapping length of two adjacent sliding windows was three sampling points. Then, we calculated the features within the time window to obtain the feature data set.

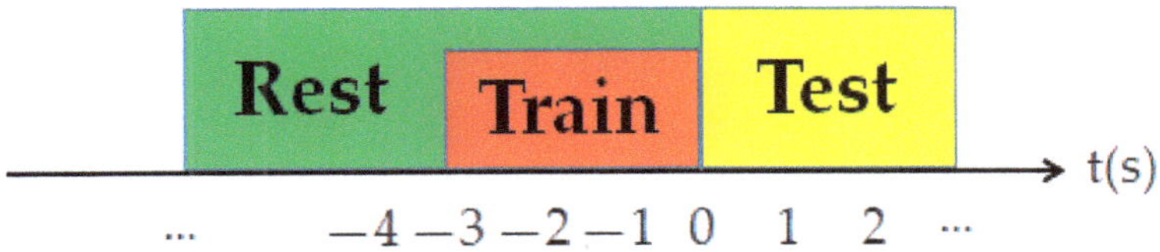

Figure 4. The schematic diagram of 4 × 4 headgear layout.

3.3. Feature Selection

3.3.1. Action Initiation Intention Feature Selection

During the data collection process, because the subjects were at rest time and the length of time in the task was different, there were also differences in the number of samples in the rest segment and the number of samples in the task segment in the collection of data. The construction of the action initiation intention recognition model faced the problem of sample imbalance between different categories. Compared with the traditional RF feature selection algorithm, this paper proposes an improved feature selection method based on RF. For the original feature set, the algorithm first set the final number of retained features, then used the sequential backward search method to remove the features with low importance score from the feature set, until the number of features was the specified number, the algorithm stopped, and the reserved feature subset was taken as the final feature set [14,15].

3.3.2. Movement State Feature Selection

Facing the high dimensionality of features in multi-classification tasks, this paper proposes a combined feature selection method based on ReliefF algorithm and genetic algorithm. In order to eliminate irrelevant features and redundant features in the original feature space, the ReliefF algorithm was first applied to evaluate the feature importance based on the distance measurement between sample features, and features with higher feature importance were retained as a feature subset [16,17]. On this basis, genetic algorithm was applied to find the optimal feature combination under the current feature subset. The flow of the feature selection method is shown in Figure 5.

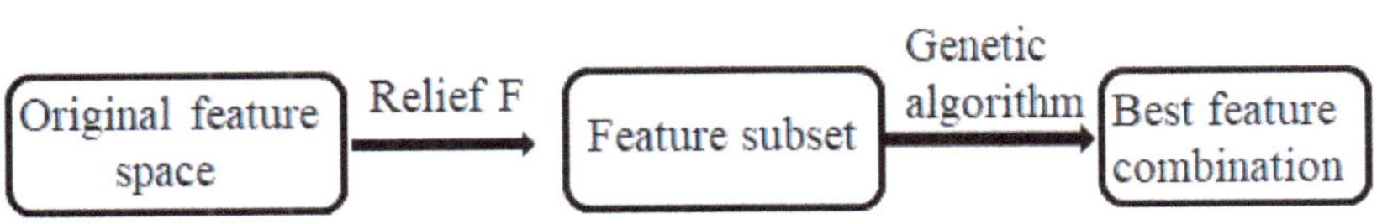

Figure 5. Feature selection flowchart.

4. Results

4.1. Data Preprocessing Results

In order to obtain the ideal morphological filtering effect, the signal-to-noise ratio and mean square error were used as evaluation indicators to conduct experiments on filters with different structural elements, different amplitudes, and different widths. We selected the parameter combination with the best comprehensive performance to construct the morphology learn filters. Table 1 shows the maximum signal-to-noise ratio SNR_{max} the minimum filtering error MSE_{min} and the corresponding combination of amplitude and width of the near-infrared signal after filtering by different structural elements.

Table 1. The filtering effect of various structural elements on the near-infrared signal.

Structural Element	Amplitude	Width	SNR_{max}	MSE_{min}
Cosine	0.8	70	11.9419	0.000219995
Triangle	1.2	60	8.19086	0.000320125
semicircle	1	70	13.0082	0.000197746
Straight line	1	50	14.9635	0.000162624

4.2. The Results and Analysis of Initial Intention Detection

4.2.1. The Results of Initial Intention of the Action

The initial intention detection task of this action divided the data samples of 50 subjects participating in the experiment into a data set, of which the experimental data of 40 subjects were used as the cross-validation set, the total number of samples was 65,354, and the experimental data of the remaining 10 subjects were used as a test set, the total number of samples was 16,392. Figure 6 shows the test results of the participants' initiation of action during the experiment. The red dashed line represented the start time of the action, and the green dashed line marked the end of the task. In the example in the figure, "0" and "1" were used to indicate the rest state and task state, respectively. If it was judged as "0" in the rest period, it was regarded as correct, and when it was judged as "1", it was regarded as a misjudgment, it was judged in the task segment. If it was "1", it was deemed correct.

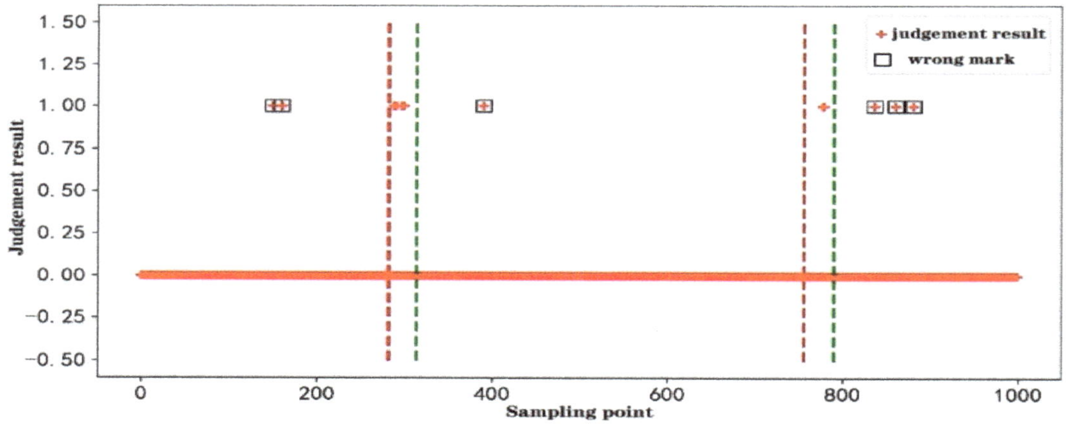

Figure 6. Schematic diagram of the judgment result of the initial intention of the action.

In order to explore the impact of different feature dimensions and different probability thresholds on the classification results, we used 40, 50, 60, 70, 80, and 90 as the feature dimensions, and applied the "threshold shift method" with 0.3, 0.4, 0.5, 0.6, and 0.7, respectively, as probability threshold to carry out the experiment. We used RF and gradient boosting tree (GBDT) classification algorithm to detect the initial intention of the action. It can be seen that when the feature dimension was 60 and the threshold was 0.4, both classification algorithms maintain their optimal performance. The optimal recognition results of the RF and GBDT are shown in Tables 2 and 3.

By comparing the comprehensive indicators of the two classifiers of RF and GBDT, the RF is finally selected as the final classification algorithm.

Table 2. Optimal results of the RF algorithm.

	Accuracy Rate	False Judgment Rate	Discrimination Delay (s)	Comprehensive Index
Train	94.4%	1.1%	−0.267	1.200
Test	92.5%	2.1%	−0.067	0.971

Table 3. Optimal results of GBDT algorithm.

	Accuracy Rate	False Judgment Rate	Discrimination Delay (s)	Comprehensive Index
Train	92.5%	1.4%	−0.202	1.113
Test	90.0%	2.5%	−0.073	0.948

4.2.2. Feature Analysis

The distribution of the brain areas of each feature and the feature weight proportion of the blood oxygen type were sorted out. The results show that among the three types of characteristics of oxygen, deoxygenation, and total oxygen, the dominant characteristic channels were channels 1, 4, and 5, and the important characteristic channels were 6, 12, 13, 19, 20 channels. The remaining channels are generally active channels, which are helpful to the recognition task.

4.3. The Results and Analysis of Exercise Status Recognition

4.3.1. Recognition Results of Motion State

In this classification task, the 50 test data samples participating in the experiment were divided into data sets. Among them, the experimental data of 40 subjects were used as the cross-validation set, the total number of samples was 160, and the experimental data of the remaining 10 subjects were used as the test set. The total number of samples was 40.

By comparing the final convergence results of the genetic algorithms of different classification algorithms, the final iterative results of the four classification algorithms are shown in Table 4. The final classification algorithm was sorted from highest to lowest score as SVM > LR > LDA > NB. Therefore, for the classification task of this data set, SVM was selected as the classifier of the final results [18]. The SVM algorithm's recognition results for all subject samples are shown in Table 5.

Table 4. The best evaluation indicators of different classifiers.

Classifier	SVM	Logistic Regression	Naive Bayes	LDA
Best Kappa coefficient	0.792	0.775	0.692	0.758

Table 5. Multi-state classification and recognition results.

	Lifting-Up	Putting Down	Pulling Back	Pushing Forward	Average	Variance	Kappa Coefficient
Train	90.0%	82.5%	82.5%	82.5%	84.4%	0.001	0.792
Test	90.0%	70.0%	90.0%	80.0%	82.5%	0.007	0.767

4.3.2. Feature Analysis

The distribution of the brain regions of each feature and the weight ratio of the blood oxygen type were classified and statistics [19]. In the characteristics of oxygen content, the dominant characteristic channels are channel 1, 4, 5, and 12, and the important characteristic channels are channel 6, 11, 13 and 15, among which channel 2, 21, 23, and 24 have no relevant characteristics, and other channels are generally active. Among the deoxygenation characteristics, the dominant characteristic channels are channel 1, 4, 5 and 6, and the important characteristic channels are channel 3, 7, 11, 12, 15, and 19. Channel 24 has no relevant characteristics, and other channels are generally active. Among the total oxygen characteristics, the dominant characteristic channels are channel 1, 4, 5, 6, 9, 12, and 13, and the important characteristic channels are channel 7, 8, 10, 17, 19, and 20. Channel 22, 23, and 24 have no relevant characteristics, while other channels are generally active.

5. Discussion

This paper proposed a set of upper limb rehabilitation training robot systems based on user spontaneous movement fNIRS-BCI. Four upper-limb movement paradigms that are highly related to daily life were designed. The combined filtering method of morphological filtering and Butterworth band-pass filtering was used to preprocess the signal. In terms of the recognition algorithm for detecting the initial intention of upper limb movements, GBDT and RF were selected for classification experiments. Finally, RF classifiers with better comprehensive indicators were selected as the final classification algorithm. The best offline recognition rate was 94.4% (151/160), the false positive rate was 1.1% (733/65354). The ReliefF algorithm based on distance measurement and the genetic algorithm proposed in the genetic theory were used to select features. In terms of upper limb motion state recognition algorithms, LR, SVM, NB, and LDA were selected for experiments, and Kappa coefficient was used as the classification index to evaluate the performance of the classifier. Finally, SVM classification got the best performance. The Kappa coefficient of the classification result was 0.792, and the four-class recognition accuracy rate was 84.4%.

This study has many advantages: First of all, in the process of the experiment, the corresponding tasks were automatically controlled by the subjects, during which there was no task hint and no external stimulation. Four upper-limb movement paradigms that are highly related to daily life were designed. At present, numerous research subjects need to carry out experiments according to stimulus cues (mostly visual or auditory stimulus). Kus et al. instructed the subjects to imagine left/right/feet movements according to the prompts displayed on the screen, and tried to classify different imaginary movements [20]. Tengfei Ma et al. instructed the subjects to perform left/right hand finger tapping task guided by a voice prompt [21]. Furthermore, in this paper, the research on upper limb movement intention was based on the blood oxygen signal before the real action task. According to the brain signal before the actual action, the motor intention was judged, which provided an important practical basis for the realization of real-time control application based on brain computer interface technology. Now the majority of research on motor imagery is based on the extraction of real motion signals for experiments, but this cannot guarantee great real-time performance. For example, Trakoolwilaiwan et al. extracted the time samples of the subjects during the resting and left and right hand movement tasks for training and recognition [22,23]. Finally, in this paper, a variety of algorithms (including ensemble learning) were selected for experiments and achieved high accuracy. In the aspect of the initial intention recognition algorithm of upper limb movement, GBDT and RF are selected for classification experiments. RF classifier with better comprehensive indicators was selected as the final classification algorithm. The best offline recognition rate was 94.4%(151/160). In the aspect of upper limb motion state recognition algorithms, LR, SVM, NB, and LDA were selected for the experiment. SVM got the highest recognition accuracy in four classification recognition (84.4%). Now, the recognition accuracy of multi-classification tasks on motor imagery is relatively low in a lot of research. Keum Shik Hong et al. simultaneously obtained fNIRS signals of mental arithmetic (MA), right hand motor imagery (RI) and left hand motor imagery (LI) from prefrontal cortex and primary motor cortex. Multiclass linear discriminant analysis was utilized to classify MA vs. RI vs. LI with an average classification accuracy of 75.6% across the ten subjects, for a 2–7 s time window during the 10 s task period [22,24]. Wang Wenle et al. collected fNIRS signals of 16 subjects' brain motor areas during the actual and imaginary movements of six types of sign language tasks. Finally, they used AdaBoost.M1, SVM, LDA, HMM, NB, and KNN algorithms to recognize the fNIRS signals. Finally, LDA achieves the highest average classification accuracy (78.70% ± 1.78%) [23,24]. The recognition accuracy in our paper is higher than 75.6% and 78.70% ± 1.78%.

Although this study has the above advantages, there are still some areas that need improvement. For example, the cascade structure adopted by the current system will cause errors to occur between different classification tasks, during which a problem in a certain recognition link will affect the final overall recognition rate. Therefore, in future research, it is expected that other real-time algorithm frameworks can be used to make up for this deficiency, such as parallel structure and hybrid structure.

Author Contributions: Conceptualization, funding acquisition, project administration, supervision, methodology, C.L.; data analysis, experimental verification, writing, Y.X.; data preprocessing, methodology, L.H.; data curation, Y.Z.; investigation, S.K.; supervision, formal analysis, L.S. All authors have read and agreed to the published version of the manuscript.

Funding: This research was funded by National Natural Science Foundation of China (61673286), and National key research and development program of China (U1713218).

Conflicts of Interest: The authors declare no conflict of interest.

References

1. Li, C.; Xu, J.; Kuang, S.; Qu, W.; Hu, H.; Sun, L. To Identify Motion Pattern of Lower Limbs by Using Cerebral Hemoglobin Information during Motor Imagery. In Proceedings of the 2017 9th International Conference on Intelligent Human-Machine Systems and Cybernetics (IHMSC), Hangzhou, China, 26–27 August 2017.
2. Myrden, A.; Chau, T. Effects of user mental state on EEG-BCI performance. *Front. Hum. Neurosci.* **2015**, *9*, 308. [CrossRef] [PubMed]
3. Tung, S.W.; Guan, C.; Kai, K.A.; Kok, S.P.; Chuan, C.W.; Ling, Z.; Wei, P.T.; Effie, C. Motor imagery BCI for upper limb stroke rehabilitation: An evaluation of the EEG recordings using coherence analysis. In Proceedings of the 2013 35th Annual International Conference of the IEEE Engineering in Medicine and Biology Society (EMBC), Osaka, Japan, 3–7 July 2013; pp. 261–264.
4. Ge, S.; Yang, Q.; Wang, R.; Lin, P.; Gao, J.; Leng, Y.; Yang, Y.; Wang, H. A Brain-Computer Interface Based on a Few-Channel EEG-fNIRS Bimodal System. *IEEE Access* **2017**, *5*, 208–218. [CrossRef]
5. Rota, G.; Handjaras, G.; Sitaram, R.; Birbaumer, N.; Dogil, G. Reorganization of functional and effective connectivity during real-time fMRI-BCI modulation of prosody processing. *Brain Lang.* **2011**, *117*, 123–132. [CrossRef] [PubMed]
6. Chowdhury, A.; Raza, H.; Meena, K.Y.; Dutta, A.; Prasad, G. An EEG-EMG correlation-based brain-computer interface for hand orthosis supported neuro-rehabilitation. *Neurosci Methods* **2019**, *15*, 1–11. [CrossRef] [PubMed]
7. Zhai, W.; Yang, Y.; Lu, S.; Gao, N. Brain computer interface system research of upper limb rehabilitation training robot. *Res. Biomed. Eng.* **2019**, *3*, 269–274.
8. Suzuki, Y.; Kaneko, N.; Sasaki, A.; Fumiya, T.; Kimitaka, N.; Taishin, N.; Matija, M. Muscle-specific movement-phase-dependent modulation of corticospinal excitability during upper-limb motor execution and motor imagery combined with virtual action observation. *Neurosci. Lett.* **2021**, *755*, 135907. [CrossRef] [PubMed]
9. Glowinski, S.; Krzyzynski, T.; Bryndal, A.; Maciejewski, I. A Kinematic Model of a Humanoid Lower LimbExoskeleton with Hydraulic Actuators. *Sensors* **2020**, *20*, 6116. [CrossRef] [PubMed]
10. Hong, K.S.; Khan, M.J.; Hong, M.J. Feature Extraction and Classification Methods for Hybrid fNIRS-EEG Brain-Computer Interfaces. *Front. Hum. Neurosci.* **2018**, *12*, 246. [CrossRef] [PubMed]
11. Tsuzuki, D.; Jurcak, V.; Ksingh, A.; Okamoto, M.; Eiju, W.; Dan, I. Virtual spatial registration of stand-alone fNIRS data to MNI space. *Neuroimage* **2007**, *34*, 1506. [CrossRef] [PubMed]
12. Saha, S.; Nesterets, Y.; Rana, R.; Tahtali, M.; Gureyev, T.E. EEG source localization using a sparsity prior based on Brodmann areas. *Int. J. Imaging Syst. Technol.* **2014**, *27*, 333–344. [CrossRef]
13. Bhateja, V.; Verma, R.; Mehrotra, R.; Urooj, S. A Non-Linear Approach to ECG Signal Processing using Morphological Filters. *Int. J. Meas. Technol. Instrum. Eng. (IJMTIE)* **2013**, *3*, 46–59. [CrossRef]
14. Jin, H.; Li, C.; Sun, L.; Hu, H.; Xu, J.; Qu, W. To classify two-dimensional motion state of step length and walking speed by applying cerebral hemoglobin information. In Proceedings of the 2017 10th International Conference on Human System Interactions (HIS), Ulsan, Korea, 17–19 July 2017.
15. Yang, L.; Song, Y.; Ma, K.; Su, E.; Xie, L. A novel motor imagery EEG decoding method based on feature separation. *J. Neural Eng.* **2021**, *18*, 036022. [CrossRef]
16. Mahmoodin, Z.; Mansor, W.; Lee, K.Y.; Mohamad, N.B. An analysis of EEG signal power spectrum density generated during writing in children with dyslexia. In Proceedings of the 2015 IEEE 11th International Colloquium on Signal Processing & its Applications (CSPA), Kuala Lumpur, Malaysia, 6–8 March 2015.
17. Wu, T.; Yan, G.; Yang, B.; Sun, H. EEG feature extraction based on wavelet packet decomposition for brain-computer interface. *Measurement* **2008**, *41*, 618–625.

18. Chaudhary, U.; Birbaumer, N.; Ramos-Murguialday, A. Brain-computer interfaces for communication and rehabilitation. *Nat. Rev. Neurol.* **2016**, *12*, 513. [CrossRef] [PubMed]
19. Durgabai, R.P.L.; Ravi, B.Y. Feature selection using ReliefF algorithm. *Int. J. Adv. Res. Comput. Commun. Eng.* **2014**, *3*, 8215–8218. [CrossRef]
20. Kus, R.; Valbuena, D.; Zygierewicz, J.; Malechka, T.; Graeser, A.; Durka, P. Asynchronous BCI Based on Motor Imagery with Automated Calibration and Neurofeedback Training. *IEEE Trans. Neural Syst. Rehabil. Eng.* **2012**, *20*, 823–835. [CrossRef] [PubMed]
21. Ma, T.; Wang, S.; Xia, Y.; Zhu, X.; Evans, J.; Sun, Y.; He, S. CNN-based classification of fNIRS signals in motor imagery BCI system. *J. Neural Eng.* **2021**, *18*, 056019. [CrossRef] [PubMed]
22. Trakoolwilaiwan, T.; Behboodi, B.; Lee, J.; Kyungson, K.; Ji-Woong, C. Convolutional neural network for high-accuracy functional near-infrared spectroscopy in a brain-computer interface: Three-class classification of rest, right-, and left-hand motor execution. *Neurophotonics* **2018**, *5*, 011008. [CrossRef] [PubMed]
23. Wang, W.; Gong, A.; Fu, Y. Identification of One—Hand Sign Language Based on fNIRS. *J. Kunming Univ. Sci. Technol. (Nat. Sci.)* **2020**, *45*, 74–81.
24. Keum-Shik, H.; Noman, N.; Yun-Hee, K. Classification of prefrontal and motor cortex signals for three-class fNIRS–BCI. *Neurosci. Lett.* **2015**, *587*, 87–92.

MDPI
St. Alban-Anlage 66
4052 Basel
Switzerland
Tel. +41 61 683 77 34
Fax +41 61 302 89 18
www.mdpi.com

Electronics Editorial Office
E-mail: electronics@mdpi.com
www.mdpi.com/journal/electronics

www.ingramcontent.com/pod-product-compliance
Lightning Source LLC
Chambersburg PA
CBHW062041220926
51850CB00010B/2624
* 9 7 8 3 0 3 6 5 5 6 7 2 7 *